SpringerBriefs in Applied Sciences and Technology

Computational Mechanics

Series Editors

Holm Altenbach⬡, Faculty of Mechanical Engineering,
Otto-von-Guericke-Universität Magdeburg, Magdeburg, Sachsen-Anhalt, Germany

Lucas F. M. da Silva, Department of Mechanical Engineering, Faculty of
Engineering, University of Porto, Porto, Portugal

Andreas Öchsner, Faculty of Mechanical Engineering, Esslingen University of
Applied Sciences, Esslingen, Germany

These SpringerBriefs publish concise summaries of cutting-edge research and practical applications on any subject of computational fluid dynamics, computational solid and structural mechanics, as well as multiphysics.

SpringerBriefs in Computational Mechanics are devoted to the publication of fundamentals and applications within the different classical engineering disciplines as well as in interdisciplinary fields that recently emerged between these areas.

More information about this subseries at http://www.springer.com/series/8886

Thomas Ranz

Linear Elasticity of Elastic Circular Inclusions Part 2/ Lineare Elastizitätstheorie bei kreisrunden elastischen Einschlüssen Teil 2

Second Edition

 Springer

Thomas Ranz
Graz, Austria

ISSN 2191-530X   ISSN 2191-5318   (electronic)
SpringerBriefs in Applied Sciences and Technology
ISSN 2191-5342   ISSN 2191-5350   (electronic)
SpringerBriefs in Computational Mechanics
ISBN 978-3-030-72396-5   ISBN 978-3-030-72397-2   (eBook)
https://doi.org/10.1007/978-3-030-72397-2

Bilingual Product

This Springer imprint is published by the registered company Springer Nature Switzerland AG
The registered company address is: Gewerbestrasse 11, 6330 Cham, Switzerland

Dedicated to my children Hannah and Lotte.
Meinen Kindern Hannah und Lotte.

Foreword

This book is a supplement to the first volume „Linear Elasticity Theory of Elastic Circular Inclusions" and now it describes the derivation of the mechanical solutions for the plane-strain-state (EVZ). In the first volume, the mechanical solutions for the plane-stress-state (ESZ) are extensively formulated, so that in this second volume a reduced formulation seems sufficient. This reduced formulation using Airy stress functions for the substructures are taken directly from the Volume 1 and therefore no repetitive derivation is necessary. Basically, with the analogy between plane-strain-state (EVZ) and plane-stress-state (ESZ), which is constituted by substitution at the constitutive level, the solution for the plane-strain-state (EVZ) can be formulated due to the already existing solution for the plane-stress-state (ESZ). In this work, substitution at the constitutive level is used exclusively for the verification of the two solutions in plane-strain-state (EVZ) and plane-stress-state (ESZ) and thus treated only secondarily. The focus of this work lies in the description of the derivation of the solution process in bilingual form.

The research work and the creation of this book took place parallel to my work at the Siemens Mobility Austria GmbH. For the good will and the unrestricted of the publication on this topic, I would like to thank Siemens Mobility Austria GmbH.

I thank Mr. B.Sc. (hon) Donald Fraser very much for the linguistic proofreading from the point of view of the native speaker.

Vorwort

Dieses Buch ist die Ergänzung zum ersten Band „Lineare Elastizitätstheorie bei kreisrunden elastischen Einschlüssen" und beschreibt die Herleitung der mechanischen Lösungen nun für den ebenen Verzerrungszustand (EVZ). Im 1. Band sind die mechanischen Lösungen für den ebenen Spannungszustand (ESZ) umfangreich formuliert, sodass in diesem 2. Band eine reduzierte Formulierung als ausreichend erscheint. Diese reduzierte Formulierung besteht darin, dass die Airyschen Spannungsfunktionen für die Teilstrukturen direkt aus dem 1. Band übernommen werden können und somit keine wiederholende Herleitung erforderlich ist. Grundsätzlich kann mit der Analogie zwischen EVZ und ESZ, welche zufolge der Substitution auf konstitutiver Ebene gebildet wird, die Lösung für den EVZ aus der bereits vorhandenen Lösung für den ESZ formuliert werden. In dieser Arbeit wird die Substitution auf konstitutiver Ebene ausschließlich für die Verifizierung der beiden Lösungen im EVZ und ESZ angewandt und somit nur sekundär behandelt. Der Fokus dieser Arbeit liegt in der Beschreibung der Herleitung des Lösungsprozesses in zweisprachiger Form.

Die Forschungstätigkeit und die Erstellung des Buches erfolgten parallel zu meiner Brotarbeit bei der Siemens Mobility Austria GmbH. Für das Entgegenkommen und die Freiheit der Publikation zu diesem Thema bedanke ich mich bei der Siemens Mobility Austria GmbH.

Herrn B.Sc. (hon) Donald Fraser danke ich besonders für die linguistische Korrekturlesung aus der Sicht des Muttersprachlers.

Graz, August 2020 Thomas Ranz

Table of contents

Inhaltsverzeichnis

Formula symbols, symbols
and abbreviations

Scalars (latin letters)

a, b	radius
p	edge load
p_i	radial parameter with index i
q_i	tangential parameter with index i
r	radial polar coordinate
u	radial displacement
v	tangential displacement
B_i, C_i, D_i	constant with index i
E	elastic modulus
F	stress function
G	shear modulus
N_1	edge load in x-direction
N_2	edge load in y-direction
N_{12}	shear edge load
S	generalization of the mechanical parameters
ΔT	difference of temperature
X_r, Y_r, Z_r	radial section-force functions
$X_\varphi, Y_\varphi, Z_\varphi$	tangential section-force functions

Formelzeichen, Symbole
und Abkürzungen

Skalare (lateinische Buchstaben)

a, b	Radius
p	Randlast
p_i	radialer Parameter mit Index i
q_i	tangentialer Parameter mit Index i
r	radiale Polarkoordinate
u	Radialverschiebung
v	Tangentialverschiebung
B_i, C_i, D_i	Konstante mit Index i
E	Elastizitätsmodul
F	Spannungsfunktion
G	Schubmodul
N_1	Randlast in x-Richtung
N_2	Randlast in y-Richtung
N_{12}	Randschublast
S	Verallgemeinerung der mechanischen Größen
ΔT	Temperaturdifferenz
X_r, Y_r, Z_r	radiale Schnittkraftfunktion
$X_\varphi, Y_\varphi, Z_\varphi$	tangentiale Schnittkraftfunktion

Scalars (greek letters)   **Skalare (griechische Buchstaben)**

α_T	coefficient of thermal expansion	α_T	Temperaturausdehnungs-koeffizient
ε_r	radial strain	ε_r	radial Dehnung
ε_φ	tangential strain	ε_φ	tangential Dehnung
$\gamma_{r\varphi}$	shear strain in $r\varphi$-plane	$\gamma_{r\varphi}$	Gleitung in der -Ebene
μ	poisson ratio	μ	Querkontraktionszahl
φ	tangential polar coordinate	φ	tangentiale Polarkoord-inate
σ_r	radial stress	σ_r	radiale Spannung
σ_φ	tangential stress	σ_φ	tangentiale Spannung
σ_1	stress load at edge in x-direction	σ_1	Spannungsrandlast in x-Richtung
σ_2	stress load at edge in y-direction	σ_2	Spannungsrandlast in y-Richtung
σ_{12}	shear stress load at edge	σ_{12}	Schubspannungsrandlast
$\tau, \sigma_{r\varphi}$	shear stress	$\tau, \sigma_{r\varphi}$	Schubspannung

Tensors (1st stage)

ε strain vector
σ stress vector

Tensors (2nd stage)

$\mathbf{J}$ compliance matrix
$\mathbf{K}$ stiffness matrix

Symbol

$\dfrac{\partial(\ )}{\partial r}$ partial derivative

Abbreviations

ESZ: means plane-stress-state

EVZ: means plane-strain-state

FEM: Finite-Element-Method

Tensoren (1. Stufe)

ε Verzerrungsvektor
σ Spannungsvektor

Tensoren (2. Stufe)

$\mathbf{J}$ Nachgiebigkeitsmatrix
$\mathbf{K}$ Steifigkeitsmatrix

Symbole

$\dfrac{\partial(\ )}{\partial r}$ partielle Ableitung

Abkürzungen

ESZ: Ebener Spannungszustand

EVZ: Ebener Verformungszustand

FEM: Finite-Elemente-Methode

Abstract

In this book the real analytic solutions for the "Disc with Circular Inclusion" under normal- and shear force at plane-strain state (EVZ) are presented. The associated solution process, which was developed according to the principle of statically indeterminate systems, is documented extensively. The solutions are given in terms of mechanical quantities (deformations, strains and stresses). Due to the superposition of the solutions for normal force in x- and y-direction and shear force the plane strain-stress relation can be formulated. The validation of the real analytic solutions is carried out by numeric FEM solution results. Comparing the results of the finite and infinite disc there is, however, a very high correspondence of all mechanical quantities. Therefore it can be assumed the real analytical solutions are the exact solutions.

The alternative formulation of the solutions for the plane-strain state (EVZ), which can be derived from the solution for the plane-strain state (ESZ) by means of substitution at the constitutive level, are shown as graphs.

Kurzfassung

In diesem Buch wird die reelle analytische Lösung für die „Scheibe mit Kern" unter Normal- und Schubkraftbelastung im ebenen Verzerrungszustand (EVZ) vorgestellt. Der zugehörige Lösungsprozess, welcher nach dem Prinzip statisch unbestimmter Systeme entwickelt wurde, wird in dieser Arbeit ausführlich dokumentiert. Als Lösungsergebnisse werden die mechanischen Größen für die Verschiebungen, Verzerrungen und Spannungen ausgewiesen. Durch die Superposition der einzelnen Lösungen, welche aus der einachsialen Belastung in x-Richtung, in y-Richtung und der Schubbelastung bestehen, lässt sich für die „Scheibe mit Kern" das ebene Elastizitätsgesetze formulieren. Die Validierung der reellen analytischen Lösungen erfolgt anhand von numerischen Ergebnissen zufolge der FEM. Trotz des Ergebnisvergleiches zwischen der unendlichen und endlichen Scheibe, der graphisch dokumentiert wird, ergibt sich eine sehr gute Übereinstimmung für alle mechanischen Größen.

Die alternative Formulierung der Lösungen für den EVZ, welche mittels der Substitution auf konstitutiver Ebene von der Lösung für den ESZ abgeleitet werden kann, wird auf der Ergebnisebene in Form von Graphen nachgewiesen.

1 Introduction

The content of this book describes in detail the derivation of the real analytic solutions for the structure "disc with circular inclusion". Thus enabling to comprehend all calculation steps. This approach was chosen to give a detailed and understandable derivation. Thus, possible arising questions can be answered independently by self-study. In no case should the detailed description create a kind of deterrence and thus prevent the application of the solutions. The interested reader will know or recognize that similar solutions to this topic are described not by so-called "one-liners". Rather, he is aware of or becomes aware that the solutions are described over several lines due to the large number of terms. A very shortened and alternative formulation of the analytic solution for the structure "disc with circular inclusion" in plane-strain-state (EVZ) is possible by the substitution at the constitutive level from the solution for the plane-stress-state (ESZ) already formulated in Volume 1 [67]. This alternative is demonstrated in this work only on the result level in the form of graphs. The focus of this work lies in the detailed description of the solution process for the "disc with circular inclusion" in bilingual form. Without extensive derivation the solutions for the "disc with ring" and "disc with inclusion and ring" from Volume 1 [67] of the plane-stress-state (ESZ) can be transferred to the plane-strain-state (EVZ)

1 Einleitung

Der Inhalt des Buches beschreibt die Herleitung der reellen analytischen Lösungen für die Struktur „Scheibe mit Kern" im Detail und ermöglicht somit die Nachvollziehung aller erforderlichen Rechenschritte. Es wurde bewusst eine sehr ausführliche Beschreibungsweise gewählt, damit die Transparenz des Lösungsprozesses gegeben ist und mögliche aufkommende Fragen selbstständig durch Eigenstudium beantwortet werden können. Auf keinen Fall sollte mit der ausführlichen Darstellung eine Art Abschreckung erzeugt und somit die Anwendung der Lösungen verhindert werden. Der interessierte Leser wird wissen bzw. erkennen, dass ähnliche Lösungen zu diesem Thema nicht durch sogenannte „Einzeiler", sondern aufgrund der großen Anzahl von Termen in den Lösungen über mehrere Zeilen beschrieben werden. Eine sehr verkürzte und alternative Formulierung der analytischen Lösung für die Struktur "Scheibe mit Kern" im EVZ ist durch die Substitution auf konstitutiver Ebene aus der bereits im 1. Band formulierten Lösung für den ESZ möglich. Diese Alternative wird in dieser Arbeit nur auf Ergebnisebene in Form von Graphen nachgewiesen. Der Fokus dieser Arbeit liegt in der ausführlichen Beschreibung des Lösungsprozesses für die "Scheibe mit Kern" in zweisprachiger Form. Ohne umfangreiche Herleitung lassen sich mit Hilfe der Substitution auf konstitutiver Ebene die Lösungen für die "Scheibe mit Ring" und "Scheibe mit Ring und

© Springer Nature Switzerland AG 2021
T. Ranz, *Linear Elasticity of Elastic Circular Inclusions Part 2/
Lineare Elastizitätstheorie bei kreisrunden elastischen Einschlüssen Teil 2*,
SpringerBriefs in Applied Sciences and Technology,
https://doi.org/10.1007/978-3-030-72397-2_1

by means of the substitution at the constitutive level.

Many tasks of strength theory and elasticity theory lead to the problem "disc with circular inclusion", which is also often referred in German to "Scheibe mit Kern" or "Scheibe mit Einschluss". For all these problems, the stress distribution or the stress transition between the disc and the inclusion is of great interest, because with a known stress distribution a simple strength evaluation on the basis of the two individual material strengths is possible. Stress increases occurring due to inclusion are exclusively caused by the change in stiffness and designated as notch stresses.

Previous papers, which will be briefly presented in section 1.2, deal with general analytical solutions to the problem „disc with circular inclusion". These general analytical solutions do not provide a complete closed solution or suggest an approximate solution. This section also presents numerical solutions (e.g., FEM) whose applications are very specialized and time consuming.

Solutions that are very close to the problem are the solutions based on the conformal mapping or those derived from complex approaches. Due to their solution characteristics or their complex approach, however, they have only limited validity at the edge between the disc and the inclusion.

Kern" aus dem 1. Band [67] vom ESZ in den EVZ überleiten.

Viele Aufgabenstellungen der Festigkeitslehre und Elastizitätstheorie führen auf das Problem „Scheibe mit Kern", welches auch gerne als „Scheibe mit Einschluss" oder im angloamerikanischen Raum als „Disc with Inclusion" bezeichnet wird. Bei all diesen Problemen ist der Spannungsverlauf bzw. Spannungsübergang zwischen Scheibe und Kern von hohem Interesse, denn bei bekanntem Spannungsverlauf ist eine einfache Festigkeitsbewertung auf Basis der beiden einzelnen Materialfestigkeiten möglich. Die durch den Einschluss auftretenden Spannungserhöhungen werden ausschließlich durch den Steifigkeitssprung hervorgerufen und als Spannungsart den Kerbspannungen zugeordnet. Bisherige Arbeiten, welche im Abschnitt 1.2 kurz vorgestellt werden, handeln von allgemeinen analytischen Lösungen für das Problem „Scheibe mit Kern", welche keine vollständige geschlossene Lösung bieten bzw. eine approximative Lösung vorschlagen, oder es werden aufwendige numerische Lösungen (z.B. FEM) vorgestellt, deren Anwendungen sehr speziell und zeitintensiv sind. Dem Problem sehr nahe kommende Lösungen stellen die auf die konforme Abbildung aufbauenden oder die mit komplexen Ansätzen hergeleiteten Lösungen dar. Sie besitzen aufgrund ihrer Lösungscharakteristik bzw. ihres komplexen Ansatzes jedoch nur eingeschränkte Gültigkeit am Rand zwischen Scheibe und Einschluss.

In this work a general real analytic solution for the "infinite disc with circular inclusion" is presented. Depending on the parameters for the load (normal force and shear) and the material (stiffness modulus and Poisson's ratio) it allows the exact analytical description of the mechanical quantities (deformations, strains and stresses) in the infinite disc and in the inclusion. For the "infinite disc with circular inclusion", the independence of the solution from the inclusion geometry (radius) is given. The solution method presented here was developed using Airy stress functions and provides the exact solution for the „plane-strain state". All formulated analytical solutions (partial and total solutions) in this work have been successfully validated by FEM analysis.

The field of application for the solution presented in this work is very broad and ranges from the microscopic level (material science) to the macroscopic level (constructive engineering). The solutions are of a fundamental nature and allow in materials science e.g. the mechanical description of material inclusions or in mechanical engineering the mechanical description of a tunnel shell or of composite materials (CFRP, GFRP, Ferro concrete, concrete). The solutions can find a significant application in the development of models for describing the fatigue strength of materials, which is particularly influenced by material inclusions.

In dieser Arbeit wird nun eine allgemeine reelle analytische Lösung für die „unendliche Scheibe mit Kern" vorgestellt. In Abhängigkeit von den Parametern für die Belastung (Normalkraft und Schub) und das Material (Steifigkeit und Querkontraktionszahl) erlaubt sie die exakte analytische Beschreibung der mechanischen Größen (Verformungen, Verzerrungen und Spannungen) in der unendlichen Scheibe und im Kern. Für die „unendliche Scheibe mit Kern" ergibt sich sinnentsprechend die Unabhängigkeit der Lösung von der Kerngeometrie (Radius). Das vorgestellte Lösungsverfahren wurde mithilfe der Airy'schen Spannungsfunktionen entwickelt und liefert für den „Ebenen Verformungszustand" (EVZ) die exakte Lösung. Alle formulierten analytischen Lösungen (Teil- und Gesamtlösungen) in dieser Arbeit sind durch eine FEM-Analyse erfolgreich validiert.

Der Anwendungsbereich für die in dieser Arbeit vorgestellte Lösung ist sehr breit gefächert und spannt sich von der mikroskopischen Ebene (Materialwissenschaft) bis zur makroskopischen Ebene (konstruktiver Ingenieurbereich). Die Lösungen sind von grundlegender Art und erlauben in der Materialwissenschaft z.B. die mechanische Beschreibung von Materialeinschlüssen oder im konstruktiven Ingenieurbereich die mechanische Beschreibung einer Tunnelschale oder von Verbundwerkstoffen (CFK, GFK, Stahlbeton, Beton). Eine wesentliche Anwendung können die Lösungen bei der Entwicklung von Modellen zur Beschreibung der Ermüdungsfestigkeit von Werkstoffen, welche besonders durch Materialeinschlüsse beeinflusst wird, finden.

1.1 Overview of chapter

In the **Chapter 1**, in addition to the introduction, there is also the literature survey for e.g. "Holes in Structures", "Inclusions", "Stress Functions", "FEM" and "Notch Stresses".

The **Chapter 2** deals with the starting point on the topic „disc with circular inclusion" and with which insights the solution process developed. The solution process is formally described therein.

Based on this in the **Chapter 3**, the real analytical solution for the disc with circular inclusion is first derived under uniaxial loading in the y-direction and then, due to transformation, the solution under uniaxial loading in the x-direction. Subsequently, the solution for the thrust load is described. The chapter concludes with the formulation of the elastic law for plane strain for the disc with circular inclusion under any normal and shear load at the infinite edge.

Validation of the real analytic solution for the „disc with circular inclusion" is done in **Chapter 4** using numerical (FEM) results. For reasons of space, only the results based on the uniaxial loading in the y-direction are taken into account. The validation of the results for the load in the x-direction and thrust loading is therefore not specifically documented in this book.

1.1 Kapitelübersicht

Im **1. Kapitel** findet sich neben der Einleitung auch der Literaturüberblick, in welchem dieser themenbezogen, für z.B. Löcher in Strukturen, Einschlüsse, Spannungsfunktionen, FEM oder Kerbspannungen, geführt wird.

Das **2. Kapitel** befasst sich mit der Ausgangslage zum Thema „Scheibe mit Kern" und anhand welcher Erkenntnisse sich der Lösungsprozess entwickelte. Der Lösungsprozess wird darin formal beschrieben.

Darauf aufbauend wird im **3. Kapitel** die reelle analytische Lösung für die Scheibe mit Kern zunächst unter einachsialer Belastung in y-Richtung und darauf aufbauend, mittels Transformation, die Lösung unter einachsialer Belastung in x-Richtung hergeleitet. Daran anschließend wird die Lösung für die Schubbelastung beschrieben. Das Kapitel schließt mit der Formulierung des ebenen Elastizitätsgesetzes für die Scheibe mit Kern unter beliebiger Normal- und Schubbelastung am unendlichen Rand.

Die Validierung der reellen analytischen Lösung für die „Scheibe mit Kern" erfolgt im **Kapitel 4** mithilfe von numerischen (FEM) Ergebnissen. Dabei werden aus Platzgründen nur die Ergebnisse zufolge der einachsialen Belastung in y-Richtung berücksichtigt. Die Validierung der Ergebnisse für die Belastung in x-Richtung und Schubbelastung wird daher in diesem Buch nicht extra dokumentiert.

A summary and the outlook for the future can be found in **Chapter 5**.

1.2 Literature survey

The literature presented here is only an extract of the literature considered as a whole in the creation process of this work and served as a stimulus for the solution process to the problem posed.

The literature search for an existing solution to the problem of „disc with circular inclusion" began with the publication by J.H. Argyris and D. Radaj „Parametrical investigation of notch stresses at an elastic core" [1]. Therein, with the finite element method, the notch stress at the elastic core is analyzed in a disc under tension. A complex analytical solution on the subject of "Infinite plane with inserted disc of other material" is given by N. J. Mußchelischwili under „Einige Grundaufgaben zur mathematischen Elastizitätstheorie" [50] or in „Some basic problems of the mathematical theory of elasticity" [49]. This solution is obtained with a complex approach and describes the problem under tensile load. This solution is not consistent with the real analytic solution described in this book, but it comes closest to the existing solutions.

Die Zusammenfassung und der Ausblick auf zukünftige Arbeiten sind im **Kapitel 5** zu finden.

1.2 Literaturüberblick

Die hier vorgestellte Literatur stellt nur einen Auszug der insgesamt beim Entstehungsprozess dieser Arbeit betrachteten Literatur dar und diente als Anregung für den Lösungsprozess zum gestellten Problem.

Die Literaturrecherche in Bezug auf eine vorhandene Lösung für das Problem „Scheibe mit Kern" begann mit der Mitteilung von J. H. Argyris und D. Radaj „Parametrische Kerbspannungsuntersuchung am elastischen Kern" [1]. Darin wird mit der Finite-Elemente-Methode die Kerbspannung am elastischen Kern in einer unter Zug stehenden Scheibe analysiert. Eine komplexe analytische Lösung zum Thema „Unendliche Ebene mit eingesetzter Kreisscheibe aus anderem Material" wird von N. J. Mußchelischwili unter „Einige Grundaufgaben zur mathematischen Elastizitätstheorie" [50] bzw. in „Some basic problems of the mathematical theory of elasticity" [49] angegeben. Diese Lösung wird mit einem komplexen Ansatz gewonnen und beschreibt das Problem unter Zugbelastung. Diese Lösung entspricht nicht der in diesem Buch beschriebenen reellen analytischen Lösung, kommt ihr jedoch von den vorhandenen Lösungen am nächsten.

Further papers dealing with the topic are „Die Scheibe mit elliptischem Kern" [30], „Über die Spannungsstörungen durch Kerben und Dellen und über die Spannungsverteilung in Verbundkörpern" [40] and „Zur Theorie der Verbundkörper" [41].

In the following a limited number of papers are cited, which serve for the preparation or contributed directly or indirectly to finding a solution.

Many papers [18], [23], [84], [85], [86], [87] deal with the topic „stress disturbances in holes" and give solutions in the form of series or tables. They provide ideas for the development of solutions for finite structures.

The problem of rigid or elastic inclusion in elliptic or circular geometry is discussed in general terms in the papers [3], [8], [11], [28], [63], [64], [77].

Fundamentals of the stress functions are found in [2], [9], [32], [49], [50], [71], [81] and, moreover, by the papers [38], [59], [74], [80], [82].

Several papers [6], [14], [17], [27], [33]; [39], [43], [47], [59], [51], [52], [54], [55], [58], [72], [78] deal with the topic of planar/spatial structures with cavities or inclusions with partial consideration of boundary layers between inclusion and matrix.

The mechanical behavior and material models for fiber composite materials is treated in the literature by a large number of papers. An introduction or detailed description in compact form

Weitere Arbeiten die sich in naher und weiterer Form mit dem Thema beschäftigen sind „Die Scheibe mit elliptischem Kern" [30], „Über die Spannungsstörungen durch Kerben und Dellen und über die Spannungsverteilung in Verbundkörpern" [40] und „Zur Theorie der Verbundkörper" [41].

Nachfolgend wird ein begrenzter Teil der Arbeiten themenbezogen zitiert, welche zur Vorbereitung gedient bzw. direkt oder indirekt zur Lösungsfindung beigetragen haben.

Viele Arbeiten [18], [23], [84], [85], [86], [87] befassen sich mit dem Thema „Spannungsstörungen bei Löchern" und geben Lösungen in Form von Reihen oder Tabellen an. Sie geben Anregungen zur Entwicklung von Lösungen für finite Strukturen.

Das Problem starrer oder elastischer Einschluss in elliptischer oder kreisrunder Geometrie wird in allgemeiner Form in den Arbeiten [3], [8], [11], [28], [63], [64], [77] behandelt.

Grundlagen zu den Spannungsfunktionen werden unter [2], [9], [32], [49], [50], [71], [81] gefunden und darüber hinaus durch die Arbeiten [38], [59], [74], [80], [82] ergänzt.

Mehrere Arbeiten [6], [14], [17], [27], [33]; [39], [43], [47], [59], [51], [52], [54], [55], [58], [72], [78] behandeln das Thema ebene/räumliche Strukturen mit Hohlräumen oder Einschlüssen unter teilweiser Berücksichtigung von Grenzschichten zwischen Einschluss und Matrix.

Das mechanische Verhalten und Materialmodelle für Faser-Verbund-Werkstoffe wird in der Literatur durch eine hohe Anzahl von Arbeiten behandelt. Eine Einführung bzw. ausführliche

give e.g. the books [5] [36], [48], [60], [76], [83] which are distinguished by the papers [7], [10], [15], [16], [19], [20], [21], [25], [29], [31], [37], [42], [45], [46], [57], [68], [73], [75], [79].

An introduction and fundamentals of finite element analysis are described in books [4] and [93]. Special FEM papers on inclusions will be dealt with in [35], [69].

The problem of notch stresses and their mathematical description is discussed extensively in books [12] and [53]. In addition, special tasks and approaches can be found in the papers [22], [24], [26], [56], [61], [62], [70], [88], [89], [90], [91], [92].

Beschreibung in kompakter Form geben z.B. die Bücher [5] [36], [48], [60], [76], [83] welche durch die Arbeiten [7], [10], [15], [16], [19], [20], [21], [25], [29], [31], [37], [42], [45], [46], [57], [68], [73], [75], [79] ergänzt werden.

Grundlagen zur Finite-Elemente-Analyse werden in den Büchern [4] und [93] beschrieben. Spezielle FEM-Arbeiten zum Thema „Einschluss" werden in [35], [69] behandelt.

Das Problem von Kerbspannungen und ihre mathematische Beschreibung wird durch die Bücher [12] und [53] sehr umfangreich erörtert. Darüber hinaus finden sich spezielle Aufgaben und Lösungsansätze in den Arbeiten [22], [24], [26], [56], [61], [62], [70], [88], [89], [90], [91], [92].

2 Initial position – Solutionprocess

Extensive numerical analyzes with varying moduli of elasticity and varying geometries (diameter and disc size) always show an analogous course of deformations, stresses and distortions at the contact area. These curves represent similar trigonometric functions but with different constant parameters. Building on this knowledge, in the first step, trigonometric section-force functions are used as statically indeterminate boundary functions. The unknown parameters of these section-force functions can be determined in the second step with the displacement compatibility at the contact area.

With the parameters known in this way, all mechanical variables (deformations, strains, stresses) in the disc and in the inclusion can be described analytically. Airy stress functions are suitable for the formulation of solutions to two-dimensional problems (ESZ, EVZ) in homogeneous material conditions. In this case, the displacement boundary conditions free and fixed, which cause a change in the continuous stress curve, can be taken into account. Airy stress functions are decoupled from constitutive relations (homogeneous materials) and therefore they have no dimensions in form of material parameters. Therefore it is not possible,

2 Ausgangslage – Lösungsprozess

Umfangreiche numerische Analysen bei variierenden E-Moduli und variierenden Geometrien (Durchmesser und Scheibengröße) zeigen an der Berührstelle stets einen analogen Verlauf der Verformungen, Spannungen und Verzerrungen. Diese Verläufe stellen gleichartige trigonometrische Funktionen dar, jedoch mit unterschiedlichen konstanten Parametern. Auf diese Erkenntnis aufbauend werden im ersten Schritt trigonometrische Schnittkraftfunktionen als statisch unbestimmte Randfunktionen herangezogen. Die unbekannten Parameter dieser Schnittkraftfunktionen können im zweiten Schritt mit der Verschiebungskompatibilität an den freigeschnittenen Rändern bestimmt werden. Mit den so bekannten Parametern können alle mechanischen Größen (Verformungen, Verzerrungen, Spannungen) in der Scheibe und im Kern analytisch beschrieben werden. Airy'sche Spannungsfunktionen eignen sich zur Formulierung von Lösungen für ebene Probleme (ESZ, EVZ) bei homogenen Materialverhältnissen. Dabei können auch die Verschiebungsrandbedingungen frei und fest, welche eine Veränderung im kontinuierlichen Spannungsverlauf verursachen, berücksichtigt werden. Die Airy'schen Spannungsfunktionen sind entkoppelt von konstitutiven Beziehungen (homogene Materialien) und besitzen daher keine Größen in Form von Materialparametern.

T. Ranz, *Linear Elasticity of Elastic Circular Inclusions Part 2/ Lineare Elastizitätstheorie bei kreisrunden elastischen Einschlüssen Teil 2*, SpringerBriefs in Applied Sciences and Technology, https://doi.org/10.1007/978-3-030-72397-2_2

in heterogeneous structures, e.g. different material stiffness, an exclusive description of the stress state of the entire structure with a single Airy stress function. This means that each individual substructure is described by a specific stress function. Heterogeneous structures represent statically indeterminate systems and can be solved by means of the constitutive relationship, in the form of displacement compatibility. The presented solutions in this book were developed under the aspect of not a series solution but to offer a closed solution. Series solutions represent approximate solutions which usually increase the degree of approximation with increasing series members. Also, due to the real solutions, the application in engineering should be given.

Daher ist bei heterogenen Strukturen, z.B. unterschiedliche Materialsteifigkeiten, eine ausschließliche Beschreibung des Spannungszustandes der Gesamtstruktur mit einer einzigen Airy'schen Spannungsfunktionen nicht möglich. Das bedeutet, dass jede einzelne Teilstruktur durch eine spezifische Spannungsfunktion beschrieben wird. Heterogene Strukturen stellen statisch unbestimmte Systeme dar und können mithilfe der konstitutiven Beziehung, in Form der Verschiebungskompatibilität, gelöst werden. Die vorgestellten Lösungen wurden unter dem Aspekt entwickelt keine Reihenlösung sondern eine geschlossene Lösung zu bieten. Reihenlösungen stellen nämlich angenäherte Lösungen dar, welche in der Regel mit zunehmenden Reihengliedern den Approximationsgrad erhöhen. Ebenso soll durch die reellen Lösungen die Anwendbarkeit im Ingenieurbereich gefördert werden.

3 Disc with circular inclusion – Plane strain condition

A practical application of the solution to this problem is given e.g. at a circular inclusion in a homogeneous material, with mechanical loading.

3.1 Unidirectional Force

This section describes the formulation of the stresses and deformations for the infinite disc with circular inclusion under uniaxial loading for the plane strain condition. For this purpose, the stress and displacement functions for the disc with hole and uniaxial force, for the disc with hole under radial and tangential load on the edge of the hole and for the circular disc, which represents the core, are formulated according to radial and tangential load on the outer edge. With the deformation functions thus obtained, which contain the unknown parameters, the compatibility is formulated at the free boundary. The displacement compatibility provides a linear system of equations for determining the unknown parameters of the section-force functions. The now determined parameters of the section-force functions then allow the

3 Scheibe mit Kern – Ebener Verformungszustand

Der praktische Anwendungsbereich der Lösung für dieses Problem ist z.B. bei einem kreisrunden Einschluss in einem homogenen Material, welches unter mechanischer Beanspruchung steht, gegeben.

3.1 Einachsiale Belastung

Dieser Abschnitt beschreibt die Formulierung der Spannungen und Verformungen für die unendliche Scheibe mit Kern unter einachsialer Belastung für den EVZ. Dazu werden die Spannungs- und Verschiebungsfunktionen für die Scheibe mit Loch zufolge einachsialer Belastung, für die Scheibe mit Loch unter Radial- und Tangentialbelastung am Lochrand und für die Kreisscheibe, welche den Kern darstellt, zufolge Radial- und Tangentialbelastung am Außenrand formuliert. Mit den so gewonnen Verschiebungsfunktionen, welche die unbekannten Parameter enthalten, wird anschließend die Kompatibilität am freigeschnittenen Rand gebildet. Die Verschiebungskompatibilität liefert ein lineares Gleichungssystem zur Bestimmung der unbekannten Parameter

© Springer Nature Switzerland AG 2021

T. Ranz, *Linear Elasticity of Elastic Circular Inclusions Part 2/ Lineare Elastizitätstheorie bei kreisrunden elastischen Einschlüssen Teil 2*, SpringerBriefs in Applied Sciences and Technology, https://doi.org/10.1007/978-3-030-72397-2_3

formulation of the stresses, strains and deformations in the disc and the inclusion.

der Schnittfunktionen. Die nun bestimmten Parameter der Schnittfunktionen erlauben sodann die Formulierung der Spannungen, Verzerrungen und Verschiebungen in der Scheibe und dem Kern.

3.1.1 Unidirectional Force σ_y

The general solution for uniaxial loading in the y-direction is formulated below. The solution for the load in the x-direction is obtained by the transformation of the solution presented here (y-direction) and described in section 3.1.10.

3.1.1 Einachsiale Last σ_y

Nachfolgend wird die allgemeine Lösung für die einachsiale Belastung in y-Richtung formuliert. Die Lösung für die Belastung in x-Richtung wird durch die Transformation der hier vorgestellten Lösung (y-Richtung) gewonnen und im Abschnitt 3.1.10 beschrieben.

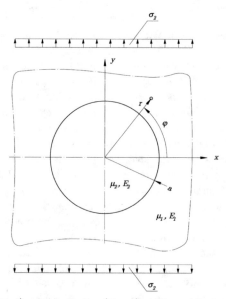

Fig. 3-1: Disc (E_1, μ_1) with inclusion (E_2, μ_2) under unidirectional loading (σ_2).

Scheibe (E_1, μ_1) mit Kern (E_2, μ_2) unter einachsialer Belastung (σ_2).

3.1.2 Procedure

The structure of the disc with circular inclusion is cut by an imaginary line at the circular contact line of inclusion and disc. The coordinate system is a polar coordinate system with the coordinates r and φ.

3.1.2 Vorgehensweise

Die Struktur der Scheibe mit Kern wird durch gedankliches Aufschneiden an der kreisrunden Berührstelle von Kern und Scheibe freigeschnitten. Als Koordinatensystem wird ein Polarkoordinatensystem mit den Koordinaten r und φ eingeführt.

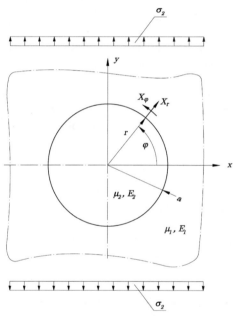

Fig. 3-2: Section-forces at the disc with circular inclusion under unidirectional force. Schnittkräfte an der Scheibe mit Kern unter einachsialer Belastung.

As unknown quantities, cutting functions in the radial and tangential direction are used as trigonometric functions. The choice of functions is made on the basis of the deformation, stress and strain states known from the numerical investigation (FEM) at the contact area between the inclusion and the disc, see Volume 1 [67]. These

Als unbekannte Größen werden Schnittfunktionen in radialer und tangentialer Richtung als trigonometrische Funktionen angesetzt. Die Wahl der Funktionen erfolgt in Anlehnung an die, aus der numerischen Untersuchung (FEM) bekannten, Verformungs-, Spannungs-, und Verzerrungszustände an der Berührstelle zwischen Kern und

states can be described by the section force functions

Scheibe, siehe 1. Band [67]. Diese Zustände können durch die Schnittkraftfunktionen

$$X_r = p_0 - p_2 \cos 2\varphi$$
$$X_\varphi = q \sin 2\varphi$$

(3.1)

The unknown parameters of these section force functions are in the radial direction p_0 and p_2, and in the tangential direction q. Their determination is made by the compatibility of the displacement on the cut-free edge. For this purpose, the tension and deformation functions for the cut disc with hole and circular disc are formulated.

vollständig beschrieben werden. Die unbekannten Parameter dieser Schnittkraftfunktionen sind in radialer Richtung p_0 und p_2, und in tangentialer Richtung q. Ihre Bestimmung erfolgt mittels der Kompatibilität der Verschiebung am freigeschnittenen Rand. Dazu werden die Spannungs- und Verformungsfunktionen für die freigeschnittene Scheibe mit Loch und Kreisscheibe formuliert.

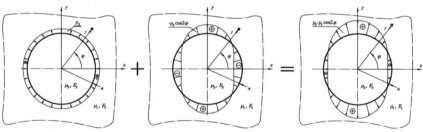

Fig. 3-3: Superposition of the unknown section functions
Superposition der unbekannten Schnittfunktionen

3.1.2.1 Extension of the solutions of the plane-stress state to the solutions of the plane-strain state

3.1.2.1 Erweiterung der Lösungen des Ebenen Spannungszustandes (ESZ) auf die Lösungen des Ebenen Verformungszustandes (EVZ)

In Volume 1 [67], the stress-describing equations of the substructures (disc, inclusion) for plane-stress state are completely described using the Airy stress functions. These stress-describing equations in the plane (r, φ)

Im 1. Band [67] sind die spannungsbeschreibenden Gleichungen der Teilstrukturen (Scheibe, Kern) für den ESZ mithilfe der Airyschen Spannungsfunktionen vollständig formuliert. Diese spannungsbeschreibenden Gleichungen

are also valid for the plane-strain state and can be taken directly from Volume 1 [67]. The stress in the z-direction is now for the plane-strain state (EVZ) not nearly zero as in the plane-stress state (ESZ). For the stress in z-direction is the following equation to taking into account.

in der Ebene (r,φ) sind ebenso für den EVZ gültig und können direkt aus dem 1. Band [67] übernommen werden. Die Spannung in z-Richtung ist nun für den EVZ nicht annähernd Null wie im ESZ sondern durch

$$\sigma_z = \mu\left(\sigma_r + \sigma_\varphi\right) \tag{3.2}$$

The shear stresses are equal to zero ($\tau_{rz} = 0$ and $\tau_{\varphi z} = 0$), as in ESZ. The relationship between the stress components in the ESZ and EVZ can also be described as follows:

zu berücksichtigen. Die Schubspannungen sind wie im ESZ gleich Null ($\tau_{rz} = 0$ und $\tau_{\varphi z} = 0$). Der Zusammenhang zwischen den Spannungskomponenten im ESZ und EVZ lässt sich auch wie folgt darstellen:

$$\begin{aligned}
\sigma_{r,ESZ} &= \sigma_{r,EVZ} \\
\sigma_{\varphi,ESZ} &= \sigma_{\varphi,EVZ} \\
\sigma_{z,ESZ} &\approx 0, \, \sigma_{z,EVZ} \neq 0 \\
\tau_{r\varphi,ESZ} &= \tau_{r\varphi,EVZ} \\
\tau_{rz,ESZ} &= 0, \, \tau_{rz,EVZ} = 0 \\
\tau_{\varphi z,ESZ} &= 0, \, \tau_{\varphi z,EVZ} = 0
\end{aligned} \tag{3.3}$$

Strains in the plane (r,φ) differ in the EVZ by the additional stress component in the z-direction compared to the ESZ. The shear strain is the same in the EVZ as in the ESZ.

Die Dehnungen in der Ebene (r,φ) unterscheiden sich im EVZ um die zusätzliche Spannungskomponente in z-Richtung gegenüber dem ESZ. Die Gleitungen sind im EVZ gleich wie im ESZ.

$$\begin{aligned}
\varepsilon_r &= \frac{1}{E}\left[\sigma_r - \mu\left(\sigma_\varphi + \sigma_z\right)\right] \\
\varepsilon_\varphi &= \frac{1}{E}\left[\sigma_\varphi - \mu\left(\sigma_r + \sigma_z\right)\right] \\
\varepsilon_z &= \frac{1}{E}\left[\sigma_z - \mu\left(\sigma_r + \sigma_\varphi\right)\right] = 0 \\
\gamma_{r\varphi} &= \frac{1}{G}\tau_{r\varphi}, \, \gamma_{rz} = 0, \, \gamma_{\varphi z} = 0
\end{aligned} \tag{3.4}$$

3.1.3 Disc with hole under unidirectional force

The infinite disc has a hole with radius *a* and is loaded in the y-direction with the stress load σ_2.

3.1.3 Scheibe mit Loch unter einachsialer Randlast

Die unendliche Scheibe besitzt ein Loch mit Radius *a* und wird in y-Richtung mit der Spannungslast σ_2 belastet.

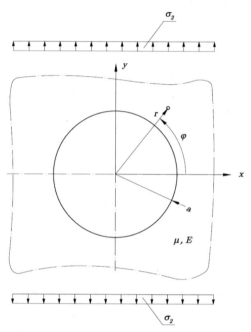

Fig. 3-4: Disc with hole under unidirectional force
Scheibe mit Loch unter einachsialer Belastung

Starting from the known kinetic relationship for the disc with hole in uniaxial loading, see, e.g. Volume 1 [67] or [9], [34], and with the completion for the tension in the z-direction

Ausgehend von der bekannten kinetischen Beziehung für die Scheibe mit Loch bei einachsialer Belastung, siehe z. B. 1. Band [67] oder [9], [34], und mit der Ergänzung für die Spannung in z-Richtung

$$\sigma_r = \frac{\sigma_2}{2}\left[1 - \frac{a^2}{r^2} - \left(1 - 4\frac{a^2}{r^2} + 3\frac{a^4}{r^4}\right)\cos 2\varphi\right]$$

$$\sigma_\varphi = \frac{\sigma_2}{2}\left[1 + \frac{a^2}{r^2} + \left(1 + 3\frac{a^4}{r^4}\right)\cos 2\varphi\right]$$

$$\sigma_z = \mu\left(\sigma_r + \sigma_\varphi\right) = \frac{\sigma_2}{2}\mu\left(2 + 4\frac{a^2}{r^2}\cos 2\varphi\right) \tag{3.5}$$

$$\tau_{r\varphi} = \frac{\sigma_2}{2}\left(1 + 2\frac{a^2}{r^2} - 3\frac{a^4}{r^4}\right)\sin 2\varphi$$

and the constitutive relationship | werden mithilfe der konstitutiven Beziehung

$$\varepsilon_r = \frac{1}{E}\left[\sigma_r - \mu\left(\sigma_\varphi + \sigma_z\right)\right]$$

$$\varepsilon_\varphi = \frac{1}{E}\left[\sigma_\varphi - \mu\left(\sigma_r + \sigma_z\right)\right] \tag{3.6}$$

$$\gamma_{r\varphi} = \frac{1}{G}\tau_{r\varphi}$$

are formulated the strains: | die Verzerrungen formuliert:

$$\varepsilon_r = \frac{\sigma_2}{2E}\left[1 - \mu - 2\mu^2 - \frac{a^2}{r^2}(1 + \mu) - \left(1 + 3\frac{a^4}{r^4} - 4\frac{a^2}{r^2}(1 - \mu)\right)(1 + \mu)\cos 2\varphi\right]$$

$$\varepsilon_\varphi = \frac{\sigma_2}{2E}\left[1 - \mu - 2\mu^2 + \frac{a^2}{r^2}(1 + \mu) + \left(1 + 3\frac{a^4}{r^4} - 4\mu\frac{a^2}{r^2}\right)(1 + \mu)\cos 2\varphi\right] \tag{3.7}$$

$$\gamma_{r\varphi} = \frac{\sigma_2}{2G}\left(1 + 2\frac{a^2}{r^2} - 3\frac{a^4}{r^4}\right)\sin 2\varphi$$

Alternatively the strains can also be formulated with reference to ESZ: | Die Verzerrungen lassen sich alternativ auch mit dem Bezug zum ESZ formulieren:

$$\varepsilon_{r,EVZ} = \varepsilon_{r,ESZ} + \frac{\sigma_2}{2E}\left[-2\mu^2 - 4\frac{a^2}{r^2}\mu^2\cos 2\varphi\right]$$

$$\varepsilon_{\varphi,EVZ} = \varepsilon_{\varphi,ESZ} + \frac{\sigma_2}{2E}\left[-2\mu^2 - 4\frac{a^2}{r^2}\mu^2\cos 2\varphi\right] \tag{3.8}$$

$$\gamma_{r\varphi,EVZ} = \gamma_{r\varphi,ESZ}$$

From the kinematic relations for the strains | Die radiale und tangentiale Verschiebung werden aus der kinematischen Beziehung für die Dehnungen

$$\varepsilon_r = \frac{\partial u}{\partial r} \quad \rightarrow \quad u = \int \varepsilon_r \mathrm{d}r = \frac{1}{E}\int (\sigma_r - \mu\sigma_\varphi)\mathrm{d}r$$

$$\varepsilon_\varphi = \frac{1}{r}\left(u + \frac{\partial v}{\partial \varphi}\right) \rightarrow \frac{\partial v}{\partial \varphi} = r\varepsilon_\varphi - u \rightarrow v = \int (r\varepsilon_\varphi - u)\mathrm{d}\varphi = \int \left[\frac{r}{E}(\sigma_\varphi - \mu\sigma_r) - u\right]\mathrm{d}\varphi$$

(3.9)

the radial and tangential displacements are obtained: gewonnen:

$$u_{EVZ} = u_{ESZ} + \frac{\sigma_2}{2E}\mu^2\left(-2r + 4\frac{a^2}{r}\cos 2\varphi\right)$$

$$= \frac{\sigma_2}{2E}\left\{ r + \frac{a^2}{r} - \left(r + 4\frac{a^2}{r} - \frac{a^4}{r^3}\right)\cos 2\varphi - \mu\left[r - \frac{a^2}{r} + \left(r - \frac{a^4}{r^3}\right)\cos 2\varphi\right] - \mu^2\left(2r - 4\frac{a^2}{r}\cos 2\varphi\right)\right\}$$

(3.10)

$$v_{EVZ} = \int (r\varepsilon_{\varphi;EVZ} - u_{EVZ})\mathrm{d}\varphi$$

$$= \int r\left[\varepsilon_{\varphi,ESZ} + \frac{\sigma_2}{2E}\left(-2\mu^2 - 4\frac{a^2}{r^2}\mu^2 \cos 2\varphi\right)\right] - \left[u_{ESZ} + \frac{\sigma_2}{2E}\mu^2\left(-2r + 4\frac{a^2}{r}\cos 2\varphi\right)\right]\mathrm{d}\varphi$$

$$= v_{ESZ} - 2\frac{\sigma_2}{E}\mu^2\frac{a^2}{r}\sin 2\varphi$$

$$= \frac{\sigma_2}{2E}\left[\left(r + \frac{a^4}{r^3}\right)(1 + \mu) + 2\frac{a^2}{r}(1 - \mu) - 4\mu^2\frac{a^2}{r}\right]\sin 2\varphi - 2\frac{\sigma_2}{E}\mu^2\frac{a^2}{r}\sin 2\varphi$$

(3.11)

The deformations on the edge of the hole thus consist of the radial displacement   Die Verformungen am Lochrand $r = a$ bestehen somit aus der radialen Verschiebung

$$u_{EVZ}(a) = \frac{\sigma_2 \cdot a}{E}(1 - \mu^2)(1 - 2\cos 2\varphi)$$

(3.12)

and the tangential displacement   und der tangentialen Verschiebung

$$v_{EVZ}(a) = \frac{2\sigma_2 \cdot a}{E}(1 - \mu^2)\sin 2\varphi$$

(3.13)

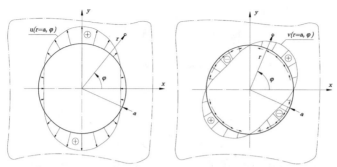

Fig. 3-5: Radial and tangential deformation at the edge of the hole at the disc
with hole.   Radial- und Tangentialverformung am Lochrand an der
Scheibe mit Loch.

3.1.4 Disc with hole under radial bearing load

The infinite disc with hole is loaded at
the bearing edge by the radial stress
load, which represents a trigonometric
function:

3.1.4 Lochscheibe unter radialer Lochleibungslast

Die unendliche Scheibe mit Loch wird
am Lochleibungsrand durch die radiale
Spannungslast, welche eine trigonometrische Funktion darstellt, belastet:

$$p(\varphi) = p_0 - p_2 \cos 2\varphi \qquad\qquad (3.14)$$

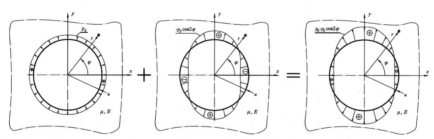

Fig. 3-6: Superposition of the radial bearing load on the disc with hole.
Superposition der radialen Lochleibungslast an der Scheibe mit Loch.

The associated stress components in
the plane (r,φ) can be taken from the
Volume 1 [67]:

Die zugehörigen Spannungskomponenten in der Ebene (r,φ) können dem
1. Band [67] entnommen werden:

$$\sigma_r = -p_0 \frac{a^2}{r^2} + \left(2p_2 \frac{a^2}{r^2} - p_2 \frac{a^4}{r^4}\right)\cos 2\varphi$$

$$\sigma_\varphi = p_0 \frac{a^2}{r^2} + p_2 \frac{a^4}{r^4}\cos 2\varphi$$

$$\sigma_z = 2\mu p_2 \frac{a^2}{r^2}\cos 2\varphi \tag{3.15}$$

$$\tau_{r\varphi} = \left(p_2 \frac{a^2}{r^2} - p_2 \frac{a^4}{r^4}\right)\sin 2\varphi$$

The constitutive relationship (3.6) can be used to formulate the strains:	Mithilfe der konstitutiven Beziehung (3.6) lassen sich die Verzerrungen formulieren:

$$\varepsilon_{r,EVZ} = \varepsilon_{r,ESZ} - 2\frac{p_2}{E}\mu^2 \frac{a^2}{r^2}\cos 2\varphi$$

$$= \frac{1}{E}\left\{-p_0 \frac{a^2}{r^2}(1+\mu) + \left[2\frac{a^2}{r^2} - \frac{a^4}{r^4}(1+\mu)\right]p_2 \cos 2\varphi\right\} - 2\frac{p_2}{E}\mu^2 \frac{a^2}{r^2}\cos 2\varphi$$

$$\varepsilon_{\varphi,EVZ} = \varepsilon_{\varphi,ESZ} - 2\frac{p_2}{E}\mu^2 \frac{a^2}{r^2}\cos 2\varphi \tag{3.16}$$

$$= \frac{1}{E}\left\{p_0 \frac{a^2}{r^2}(1+\mu) + \left[-2\mu\frac{a^2}{r^2} + \frac{a^4}{r^4}(1+\mu)\right]p_2 \cos 2\varphi\right\} - 2\frac{p_2}{E}\mu^2 \frac{a^2}{r^2}\cos 2\varphi$$

$$\gamma_{r\varphi,EVZ} = \gamma_{r\varphi,ESZ}$$

$$= \frac{p_2}{G}\left(\frac{a^2}{r^2} - \frac{a^4}{r^4}\right)\sin 2\varphi$$

The radial and tangential displacements are obtained from the kinematic relationship for strains (3.9):	Die radiale und tangentiale Verschiebung werden aus der kinematischen Beziehung für die Dehnungen (3.9) gewonnen:

$$u_{EVZ} = u_{ESZ} + 2\frac{p_2}{E}\mu^2 \frac{a^2}{r}\cos 2\varphi$$

$$= \frac{1}{E}\left[p_0 \frac{a^2}{r}(1+\mu) + p_2 \frac{a^4}{3r^3}(1+\mu)\cos 2\varphi - 2p_2 \frac{a^2}{r}\cos 2\varphi\right] + 2\frac{p_2}{E}\mu^2 \frac{a^2}{r}\cos 2\varphi \tag{3.17}$$

$$v_{EVZ} = v_{ESZ} - 2\frac{p_2}{E}\mu^2 \frac{a^2}{r}\sin 2\varphi$$

$$= \frac{1}{E}\left[p_2 \frac{1}{3}\frac{a^4}{r^3}(1+\mu) + p_2 \frac{a^2}{r}(1-\mu)\right]\sin 2\varphi - 2\frac{p_2}{E}\mu^2 \frac{a^2}{r}\sin 2\varphi \tag{3.18}$$

The deformations at the edge of the hole $r = a$ can thus be described by the radial displacement

Die Verformungen am Lochrand $r = a$ können somit beschrieben werden durch die radiale Verschiebung

$$u_{EVZ}(a) = \frac{a}{E}\left[p_0(1+\mu) + \frac{p_2}{3}(\mu-5)\cos 2\varphi \right] + 2\frac{p_2}{E}\mu^2 a\cos 2\varphi \qquad (3.19)$$

and tangential displacement:

und die tangentiale Verschiebung:

$$v_{EVZ}(a) = \frac{2p_2 \cdot a}{3E}(2-\mu)\sin 2\varphi - 2\frac{p_2}{E}\mu^2 a\sin 2\varphi \qquad (3.20)$$

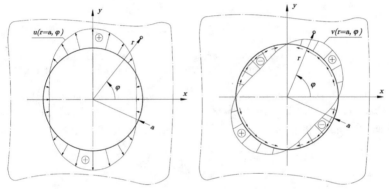

Fig. 3-7: Radial and tangential deformation at the hole edge on the disc with hole under radial bearing load. Radial- und Tangentialverformung am Lochrand an der Scheibe mit Loch unter radialer Lochleibungslast.

3.1.5 Disc with hole under tangential bearing load

3.1.5 Lochscheibe mit tangentialer Lochleibungslast

The infinite disc with hole is loaded at the bearing edge by the tangential stress load, which represents a trigonometric function:

Die unendliche Scheibe mit Loch wird am Lochleibungsrand durch die tangentiale Spannungslast, welche ebenso eine trigonometrische Funktion darstellt, belastet:

$$q(\varphi) = q\sin 2\varphi \qquad (3.21)$$

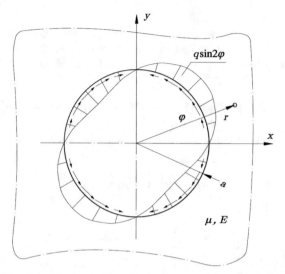

Fig. 3-8: Tangential bearing load on the disc with hole.
Tangentiale Lochleibungslast an der Scheibe mit Loch.

The associated stress components in the plane (r,φ) can be taken again from the Volume 1 [67]:

Die zugehörigen Spannungskomponenten in der Ebene (r,φ) können wieder dem 1. Band [67] entnommen werden:

$$\sigma_r = 2q\left(\frac{a^2}{r^2} - \frac{a^4}{r^4}\right)\cos 2\varphi$$

$$\sigma_\varphi = 2q\frac{a^4}{r^4}\cos 2\varphi$$

$$\sigma_z = 2q\mu\frac{a^2}{r^2}\cos 2\varphi \qquad (3.22)$$

$$\tau_{r\varphi} = 2q\left(\frac{a^2}{2r^2} - \frac{a^4}{r^4}\right)\sin 2\varphi$$

The constitutive relationship (3.6) can be used to formulate the strains:

Mit der konstitutiven Beziehung (3.6) können die Verzerrungen formuliert werden:

$$\varepsilon_{r,EVZ} = \varepsilon_{r,ESZ} - 2\frac{q}{E}\mu^2\frac{a^2}{r^2}\cos 2\varphi$$

$$= \frac{2q}{E}\left[\frac{a^2}{r^2} - \frac{a^4}{r^4}(1+\mu)\right]\cos 2\varphi - 2\frac{q}{E}\mu^2\frac{a^2}{r^2}\cos 2\varphi$$

$$\varepsilon_{\varphi,EVZ} = \varepsilon_{\varphi,ESZ} - 2\frac{q}{E}\mu^2\frac{a^2}{r^2}\cos 2\varphi$$

$$= \frac{2q}{E}\left[-\mu\frac{a^2}{r^2} + \frac{a^4}{r^4}(1+\mu)\right]\cos 2\varphi - 2\frac{q}{E}\mu^2\frac{a^2}{r^2}\cos 2\varphi$$

$$\gamma_{r\varphi,EVZ} = \gamma_{r\varphi,ESZ}$$

$$= \frac{2q}{G}\left(\frac{a^2}{2r^2} - \frac{a^4}{r^4}\right)\sin 2\varphi$$

(3.23)

The radial and tangential displacements are obtained from the kinematic relationship for strains (3.9):

Die radiale und tangentiale Verschiebung werden aus der kinematischen Beziehung für die Dehnungen (3.9) gewonnen.

$$u_{EVZ} = u_{ESZ} + 2\frac{q}{E}\mu^2\frac{a^2}{r}\cos 2\varphi$$

$$= \frac{2q}{3E}\left[-3\frac{a^2}{r} + (1+\mu)\frac{a^4}{r^3}\right]\cos 2\varphi + 2\frac{q}{E}\mu^2\frac{a^2}{r}\cos 2\varphi$$

(3.24)

$$v_{EVZ} = v_{ESZ} - 2\frac{q}{E}\mu^2\frac{a^2}{r}\sin 2\varphi$$

$$= \frac{q}{3E}\left[3(1-\mu)\frac{a^2}{r} + 2(1+\mu)\frac{a^4}{r^3}\right]\sin 2\varphi - 2\frac{q}{E}\mu^2\frac{a^2}{r}\sin 2\varphi$$

(3.25)

The deformations at the edge of the hole $r = a$ can thus be described by the radial displacement

Die Verformungen am Lochrand $r = a$ können somit durch die radiale Verschiebung

$$u(a) = \frac{2}{3}\frac{a \cdot q}{E}(\mu - 2)\cos 2\varphi + 2\frac{q}{E}\mu^2 a\cos 2\varphi$$

(3.26)

and tangential displacement:

und die tangentiale Verschiebung beschrieben werden:

$$v(a) = \frac{q \cdot a}{3E}(5 - \mu)\sin 2\varphi - 2\frac{q}{E}\mu^2 a\sin 2\varphi$$

(3.27)

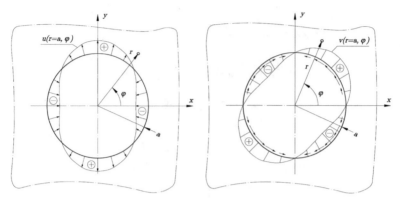

Fig. 3-9: Radial and tangential deformation at the hole edge on the disc with hole under tangential bearing load. Radial- und Tangentialverformung am Lochrand an der Scheibe mit Loch unter tangentialer Lochleibungslast.

3.1.6 Circular disc under radial edge load

The circular disc is loaded at the outer edge by the radial stress load, which represents a trigonometric function:

3.1.6 Kreisscheibe unter radialer Randlast

Die Kreisscheibe wird am Außenrand durch die radiale Spannungslast, welche eine trigonometrische Funktion darstellt, belastet:

$$p(\varphi) = -p_0 + p_2 \cos 2\varphi \ , \tag{3.28}$$

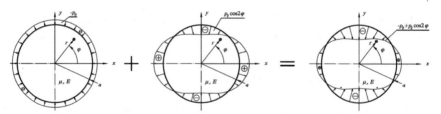

Fig. 3-10: Superposition of the radial edge load on the circular disc. Superposition der radialen Randlast an der Kreisscheibe.

The associated stress components in the plane (r,φ) can be taken again from the Volume 1 [67]:

Die zugehörigen Spannungskomponenten in der Ebene (r,φ) können wieder dem 1. Band [67] entnommen werden:

$$\sigma_r = -p_0 + p_2 \cos 2\varphi$$

$$\sigma_\varphi = -p_0 + \left(-p_2 + 2p_2 \frac{r^2}{a^2}\right)\cos 2\varphi$$

$$\sigma_z = \mu\left(-2p_0 + 2p_2 \frac{r^2}{a^2}\cos 2\varphi\right) \tag{3.29}$$

$$\tau_{r\varphi} = \left(-p_2 + p_2 \frac{r^2}{a^2}\right)\sin 2\varphi$$

The constitutive relationship (3.6) provides for the strains:

Die konstitutive Beziehung (3.6) liefert für die Verzerrungen:

$$\varepsilon_{r,EVZ} = \varepsilon_{r,ESZ} - \frac{1}{E}\mu^2\left(-2p_0 + 2p_2 \frac{r^2}{a^2}\cos 2\varphi\right)$$

$$= \frac{1}{E}\left[p_0(\mu-1) + \left(1 + \mu - 2\mu\frac{r^2}{a^2}\right)p_2 \cos 2\varphi\right] - \frac{1}{E}\mu^2\left(-2p_0 + 2p_2 \frac{r^2}{a^2}\cos 2\varphi\right)$$

$$\varepsilon_{\varphi,EVZ} = \varepsilon_{\varphi,ESZ} - \frac{1}{E}\mu^2\left(-2p_0 + 2p_2 \frac{r^2}{a^2}\cos 2\varphi\right) \tag{3.30}$$

$$= \frac{1}{E}\left[p_0(\mu-1) + \left(-1 - \mu + 2\frac{r^2}{a^2}\right)p_2 \cos 2\varphi\right] - \frac{1}{E}\mu^2\left(-2p_0 + 2p_2 \frac{r^2}{a^2}\cos 2\varphi\right)$$

$$\gamma_{r\varphi,EVZ} = \gamma_{r\varphi,ESZ}$$

$$= \frac{p_2}{G}\left(-1 + \frac{r^2}{a^2}\right)\sin 2\varphi$$

The displacements are: Die Verschiebungen lauten:

$$u_{EVZ} = u_{ESZ} - \frac{2}{E}\mu^2\left(-p_0 r + p_2 \frac{r^3}{3a^2}\cos 2\varphi\right)$$

$$= \frac{1}{E}\left\{-p_0(1-\mu)r + p_2\left[(1+\mu)r - \mu\frac{2r^3}{3a^2}\right]\cos 2\varphi\right\} - \frac{2}{E}\mu^2\left(-p_0 r + p_2 \frac{r^3}{3a^2}\cos 2\varphi\right)$$

$$(3.31)$$

$$v_{EVZ} = v_{ESZ} - \frac{2}{E}\mu^2 p_2 \frac{r^3}{3a^2}\sin 2\varphi$$

$$(3.32)$$

$$= \frac{1}{E}\left[-r(1+\mu) + \frac{1}{3}\frac{r^3}{a^2}(3+\mu)\right]p_2\sin 2\varphi - \frac{2}{E}\mu^2 p_2 \frac{r^3}{3a^2}\sin 2\varphi$$

The deformations at the outer edge $r = a$ can thus be described by the radial displacement

Die Verformungen am Außenrand $r = a$ können somit durch die radiale Verschiebung

$$u_{EVZ}(a) = \frac{1}{E}\left[-p_0(1-\mu)a + \frac{a}{3}p_2(3+\mu)\cos 2\varphi\right] - \frac{2}{E}\mu^2\left(-p_0 a + p_2 \frac{a}{3}\cos 2\varphi\right) \quad (3.33)$$

and tangential displacement: und die tangentiale Verschiebung beschrieben werden:

$$v_{EVZ}(a) = -\frac{2p_2 \cdot a}{3E}\mu\sin 2\varphi - \frac{2}{E}\mu^2 p_2 \frac{a}{3}\sin 2\varphi \quad (3.34)$$

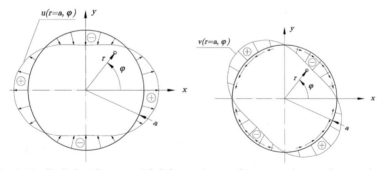

Fig. 3-11: Radial and tangential deformation at the outer edge on the circular disc under radial edge load. Radial- und Tangentialverformung am Kreisscheibenrand unter radialer Randlast.

3.1.7 Circular disc under tangential edge load

The circular disc is loaded at the outer edge by the tangential stress load:

3.1.7 Kreisscheibe unter tangentialer Randlast

Die Kreisscheibe wird am Rand durch die tangentiale Spannungslast belastet:

$$q(\varphi) = -q\sin 2\varphi \qquad (3.35)$$

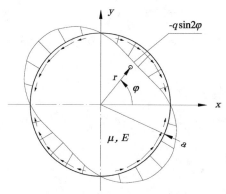

Fig. 3-12: Tangential edge load on the circular disc.
Tangentiale Randlast an der Kreisscheibe.

The associated stress components in the plane (r,φ) can be taken again from the Volume 1 [67]:

Die zugehörigen Spannungskomponenten in der Ebene (r,φ) können wieder dem 1. Band [67] entnommen werden:

$$\sigma_r = 0$$

$$\sigma_\varphi = -2q\frac{r^2}{a^2}\cos 2\varphi$$

$$\sigma_z = -2q\mu\frac{r^2}{a^2}\cos 2\varphi \qquad (3.36)$$

$$\tau_{r\varphi} = -q\frac{r^2}{a^2}\sin 2\varphi$$

The constitutive relationship (3.6) provides for the strains:

Mithilfe der konstitutiven Beziehung (3.6) folgt für die Verzerrungen:

$$\varepsilon_{r,EVZ} = \varepsilon_{r,ESZ} + \frac{2q}{E}\mu^2 \frac{r^2}{a^2}\cos 2\varphi$$

$$= \frac{2q}{E}\mu \frac{r^2}{a^2}\cos 2\varphi + \frac{2q}{E}\mu^2 \frac{r^2}{a^2}\cos 2\varphi$$

$$\varepsilon_{\varphi,EVZ} = \varepsilon_{\varphi,ESZ} + \frac{2q}{E}\mu^2 \frac{r^2}{a^2}\cos 2\varphi \qquad\qquad (3.37)$$

$$= -\frac{2q}{E}\frac{r^2}{a^2}\cos 2\varphi + \frac{2q}{E}\mu^2 \frac{r^2}{a^2}\cos 2\varphi$$

$$\gamma_{r\varphi,EVZ} = \gamma_{r\varphi,ESZ}$$

$$= -\frac{q}{G}\frac{r^2}{a^2}\sin 2\varphi$$

The displacements are: Die Verschiebungen lauten:

$$u_{EVZ} = u_{ESZ} + \frac{2q}{3E}\mu^2 \frac{r^3}{a^2}\cos 2\varphi$$
$$\qquad\qquad\qquad\qquad\qquad\qquad\qquad (3.38)$$
$$= \frac{2q}{3E}\mu \frac{r^3}{a^2}\cos 2\varphi + \frac{2q}{3E}\mu^2 \frac{r^3}{a^2}\cos 2\varphi = \frac{2q}{3E}\frac{r^3}{a^2}\mu(1+\mu)\cos 2\varphi$$

$$v_{EVZ} = v_{EVZ} + \frac{2q}{3E}\mu^2 \frac{r^3}{a^2}\sin 2\varphi$$
$$\qquad\qquad\qquad\qquad\qquad\qquad\qquad (3.39)$$
$$= -\frac{q}{3E}\frac{r^3}{a^2}(3+\mu)\sin 2\varphi + \frac{2q}{3E}\mu^2 \frac{r^3}{a^2}\sin 2\varphi = \frac{q}{3E}(2\mu^2-\mu-3)\frac{r^3}{a^2}\sin 2\varphi$$

The deformations at the outer edge $r = a$ can thus be described by the radial displacement

Die Verformungen am Lochrand $r = a$ können somit beschrieben werden durch die radiale Verschiebung

$$u_{EVZ}(a) = \frac{2q}{3E}a\mu(1+\mu)\cos 2\varphi \qquad\qquad (3.40)$$

and tangential displacement: und die tangentiale Verschiebung:

$$v_{EVZ}(a) = \frac{q}{3E}a(2\mu^2-\mu-3)\sin 2\varphi \qquad\qquad (3.41)$$

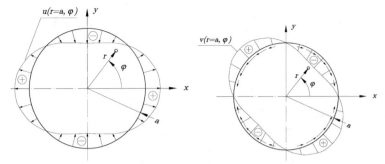

Fig. 3-13: Radial and tangential deformation at the outer edge on the circular
disc under tangential edge load. Radial- und Tangentialverformung
am Kreisscheibenrand unter tangentialer Randlast.

Hence all the required edge deformations are known and the compatibility for determining the statically indeterminate parameters (q, p_0, p_2) are determinated.

Damit sind nun alle erforderlichen Randverformungen bekannt und es kann die Kompatibilität zur Bestimmung der statisch unbestimmten Größen (q, p_0, p_2) durchgeführt werden.

3.1.8 Compatibility

3.1.8 Kompatibilität

With the well-known displacement functions at the cut-free edge between the disc and the inclusion, the displacement compatibility can now be used to determine the statically indeterminate quantities. This is based on the deformation of the disc with hole, which is now reduced with the statically indeterminate sizes. This resulting reduced displacement in the disc with hole corresponds to the displacement of the circular disc due to the statically indeterminate quantities. It should be noted that the displacements are applied in correct direction. The following illustrations show these displacements in the correct direction on the cut-free system.

Mit den bekannten Verschiebungsfunktionen an der freigeschnittenen Berandung zwischen Scheibe und Kern kann nun, zur Bestimmung der statisch unbestimmten Größen, die Verschiebungskompatibilität angewandt werden. Dazu wird von der Verformung an der Scheibe mit Loch ausgegangen, welche nun mit den statisch unbestimmten Größen reduziert wird. Diese sich dadurch reduzierte Verschiebung in der Scheibe entspricht der Verschiebung der Kernscheibe zufolge der statisch unbestimmten Größen. Zu beachten ist, dass die Verschiebungen richtungstreu angesetzt werden. Die nachfolgenden Abbildungen zeigen diese richtungstreuen Verschiebungen am freigeschnittenen System.

Disc with hole under unidirectional
load σ_2:

Scheibe mit Loch unter einachsialer
Randlast σ_2:

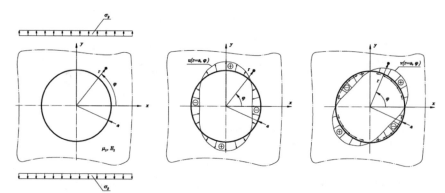

Fig. 3-14: Uniaxial load and deformations $u(r = a, \varphi)$, $v(r = a, \varphi)$, at the edge
of the hole on the disc with hole. Einachsiale Last σ_2 und Verfor
mungen $u(r = a, \varphi)$, $v(r = a, \varphi)$ am Lochrand der Scheibe mit Loch.

Radial deformation:

Radiale Verschiebung:

$$u_{EVZ}(a) = \frac{\sigma_2 \cdot a}{E_1}\left(1 - \mu_1^2\right)\left(1 - 2\cos 2\varphi\right) = u_0 \tag{3.42}$$

Tangential deformation:

Tangentiale Verschiebung:

$$v_{EVZ}(a) = \frac{2\sigma_2 \cdot a}{E_1}\left(1 - \mu_1^2\right)\sin 2\varphi = v_0 \tag{3.43}$$

Disc with hole under radial bearing
load $X_r(\varphi) = -p_0 + p_2 \cos 2\varphi$:

Scheibe mit Loch unter radialer Lochleibungslast $X_r(\varphi) = -p_0 + p_2 \cos 2\varphi$:

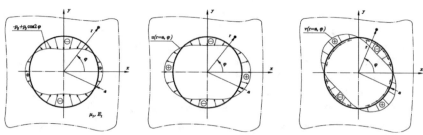

Fig. 3-15: Radial bearing load $X_r(\varphi) = -p_0 + p_2 \cos 2\varphi$ and deformations
$u(r = a, \varphi)$, $v(r = a, \varphi)$, at the edge of the hole on the disc with
hole. Radiale Lochleibungslast $X_r(\varphi) = -p_0 + p_2 \cos 2\varphi$ und Ver
formungen $u(r = a, \varphi)$, $v(r = a, \varphi)$ am Lochrand der Scheibe mit
Loch.

Radial deformation:

Radiale Verschiebung:

$$u_{EVZ}(a) = -\frac{a}{E_1}\left[p_0(1+\mu)+\frac{p_2}{3}(\mu_1-5)\cos 2\varphi\right]-2\frac{p_2}{E_1}a\mu_1^2\cos 2\varphi = u_{S,X_r} \qquad (3.44)$$

Tangential deformation:

Tangentiale Verschiebung:

$$v_{EVZ}(a) = -\frac{2p_2}{3E_1}a(2-\mu_1)\sin 2\varphi + 2\frac{p_2}{E_1}a\mu_1^2\sin 2\varphi = v_{S,X_r} \qquad (3.45)$$

Disc with hole under tangential bearing load $X_\varphi(\varphi) = -q\sin 2\varphi$:

Scheibe mit Loch unter tangentialer Lochleibungslast $X_\varphi(\varphi) = -q\sin 2\varphi$:

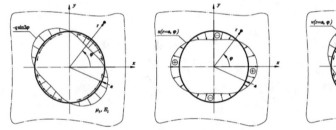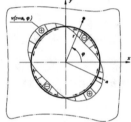

Fig. 3-16: Tangential bearing load $X_\varphi(\varphi) = -q\sin 2\varphi$ and deformations
$u(r = a,\varphi)$, $v(r = a,\varphi)$ at the edge of the hole on the disc with hole.
Tangentiale Lochleibungslast $X_\varphi(\varphi) = -q\sin 2\varphi$ und Verformungen
$u(r = a,\varphi)$, $v(r = a,\varphi)$ am Lochrand an der Scheibe mit Loch.

Radial deformation:

Radiale Verschiebung:

$$u_{EVZ}(a) = \frac{2}{3}\frac{q}{E_1}a(2-\mu_1)\cos 2\varphi - 2\frac{q}{E_1}a\mu_1^2\cos 2\varphi = u_{S,X_\varphi} \qquad (3.46)$$

Tangential deformation:

Tangentiale Verschiebung:

$$v_{EVZ}(a) = \frac{q}{3E_1}a(\mu_1-5)\sin 2\varphi + 2\frac{q}{E_1}a\mu_1^2\sin 2\varphi = v_{S,X_\varphi} \qquad (3.47)$$

Circular disc under radial bearing load
$X_r(\varphi) = p_0 - p_2 \cos 2\varphi$:

Kreisscheibe unter radialer Randlast
$X_r(\varphi) = p_0 - p_2 \cos 2\varphi$:

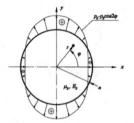
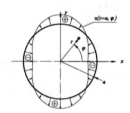
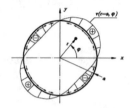

Fig. 3-17: Radial edge load $X_r(\varphi) = p_0 - p_2 \cos 2\varphi$ and deformations
$u(r = a, \varphi)$, $v(r = a, \varphi)$ at the circular disc edge. Radiale Randlast
$X_r(\varphi) = p_0 - p_2 \cos 2\varphi$ und Verformungen $u(r = a, \varphi)$, $v(r = a, \varphi)$
am Kreisscheibenrand.

Radial deformation: Radiale Verschiebung:

$$u_{EVZ}(a) = \frac{a}{E_2}\left[p_0(1-\mu_2) - \frac{p_2}{3}(3+\mu_2)\cos 2\varphi \right] + \frac{2a}{E_2}\mu_2^2\left(-p_0 + \frac{p_2}{3}\cos 2\varphi\right) = u_{K,X_r} \quad (3.48)$$

Tangential deformation: Tangentiale Verschiebung:

$$v_{EVZ}(a) = \frac{2p_2}{3E_2}a\mu_2\sin 2\varphi + \frac{2p_2}{3E_2}a\mu_2^2\sin 2\varphi = v_{K,X_r} \quad (3.49)$$

Circular disc under tangential bearing
load $X_\varphi(\varphi) = q\sin 2\varphi$:

Kreisscheibe unter tangentialer Randlast $X_\varphi(\varphi) = q\sin 2\varphi$:

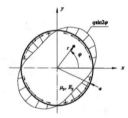
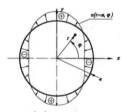
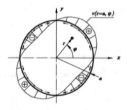

Fig. 3-18: Tangential edge load $X_\varphi(\varphi) = q\sin 2\varphi$ and deformations $u(r = a, \varphi)$,
$v(r = a, \varphi)$ at the circular disc edge. Tangentiale Randlast
$X_\varphi(\varphi) = q\sin 2\varphi$ und Verformungen $u(r = a, \varphi)$, $v(r = a, \varphi)$ am
Kreisscheibenrand.

Radial deformation: Radiale Verschiebung:

$$u_{EVZ}(a) = -\frac{2q}{3E_2}a\mu_2\left(1+\mu_2\right)\cos 2\varphi = u_{K,X_\varphi} \qquad (3.50)$$

Tangential deformation: Tangentiale Verschiebung:

$$v_{EVZ}(a) = -\frac{q}{3E_2}a\left(2\mu_2^2 - \mu_2 - 3\right)\sin 2\varphi = v_{K,X_\varphi} \qquad (3.51)$$

Compatibility of displacement: **Verschiebungskompatibilität:**

The compatibility of displacement is   Die Verschiebungskompatibilität lautet
given by: somit

$$u_0 + \left(u_{S,X_r} + u_{S,X_\varphi}\right) = u_{K,X_r} + u_{K,X_\varphi}$$
$$v_0 + \left(v_{S,X_r} + v_{S,X_\varphi}\right) = v_{K,X_r} + v_{K,X_\varphi} \qquad (3.52)$$

It equates the displacements in the disc   und setzt die Verschiebungen in der
and the inclusion at the common edge.   Scheibe und dem Kern gleich. Sie lie-
It provides two equations for the three   fert zwei Gleichungen für die insgesamt
unknown parameters $\left(q, p_0, p_2\right)$ of the   drei unbekannten Parameter $\left(q, p_0, p_2\right)$
section force functions X_r, X_φ. An ad-   der   Schnittkraftfunktionen X_r, X_φ.
ditionally required equation for the   Eine zusätzlich erforderliche Gleichung
third parameter is obtained with the   für den dritten Parameter wird mit der
condition that the first derivation of   Bedingung, dass auch die 1. Ableitung
the displacement function $\dfrac{\partial u}{\partial \varphi}$ must be   der Verschiebungsfunktion $\dfrac{\partial u}{\partial \varphi}$ erfüllt
also satisfied.   sein muss, gewonnen.

$$\frac{\partial}{\partial \varphi}\left[u_0(\varphi) + \left(u_{S,X_r}(\varphi) + u_{S,X_\varphi}(\varphi)\right) = u_{K,X_r}(\varphi) + u_{K,X_\varphi}(\varphi)\right] \qquad (3.53)$$

Consideration of first derivation $\dfrac{\partial u}{\partial \varphi}$:   Betrachtung der 1. Ableitung $\dfrac{\partial u}{\partial \varphi}$:

$$4\frac{\sigma_2 \cdot a}{E_1}\left(1-\mu_1^2\right)\sin 2\varphi + \frac{2a}{3E_1}p_2\left(\mu_1 - 5\right)\sin 2\varphi + 4p_2 a\frac{\mu_1^2}{E_1}\sin 2\varphi - \frac{4a \cdot q}{3E_1}\left(2-\mu_1\right)\sin 2\varphi$$

$$+4qa\frac{\mu_1^2}{E_1}\sin 2\varphi = \frac{2a}{3E_2}p_2\left(3+\mu_2\right)\sin 2\varphi - 4p_2\frac{a}{3}\frac{\mu_2^2}{E_2}\sin 2\varphi + \frac{4q \cdot a}{3E_2}\mu_2 \sin 2\varphi + 4q\frac{a}{3}\frac{\mu_2^2}{E_2}\sin 2\varphi$$

$$\qquad (3.54)$$

$$\frac{\sigma_2}{E_1}\left(1-\mu_1^2\right)+\frac{p_2}{3E_1}\left(\mu_1-5+6\mu_1^2\right)+\frac{2}{3}\frac{q}{E_1}\left(\mu_1-2+3\mu_1^2\right)=\frac{p_2}{3E_2}\left(3+\mu_2-2\mu_2^2\right)+\frac{2}{3}\frac{q}{E_2}\left(\mu_2+\mu_2^2\right)$$

$$p_2\left[\frac{E_2\left(\mu_1-5+6\mu_1^2\right)-E_1\left(3+\mu_2-2\mu_2^2\right)}{3E_1E_2}\right]=q\left[\frac{2E_1\left(\mu_2+\mu_2^2\right)-2E_2\left(\mu_1-2+3\mu_1^2\right)}{3E_1E_2}\right]-\sigma_2\frac{6E_2\left(1-\mu_1^2\right)}{3E_1E_2}$$

$$p_2=q\underbrace{\left[\frac{2E_1\left(\mu_2+\mu_2^2\right)-2E_2\left(\mu_1-2+3\mu_1^2\right)}{E_2\left(\mu_1-5+6\mu_1^2\right)-E_1\left(3+\mu_2-2\mu_2^2\right)}\right]}_{A_{EVZ}}-\sigma_2\underbrace{\frac{6E_2\left(1-\mu_1^2\right)}{E_2\left(\mu_1-5+6\mu_1^2\right)-E_1\left(3+\mu_2-2\mu_2^2\right)}}_{B_{EVZ}}$$

$$(3.55)$$

With the abbreviations   Mit den Abkürzungen

$$A_{EVZ}=\left[\frac{2E_1\left(\mu_2+\mu_2^2\right)-2E_2\left(\mu_1-2+3\mu_1^2\right)}{E_2\left(\mu_1-5+6\mu_1^2\right)-E_1\left(3+\mu_2-2\mu_2^2\right)}\right],\quad B_{EVZ}=\frac{6E_2\left(1-\mu_1^2\right)}{E_2\left(\mu_1-5+6\mu_1^2\right)-E_1\left(3+\mu_2-2\mu_2^2\right)}$$

$$(3.56)$$

follows: folgt:

$$p_2=q\cdot A_{EVZ}-\sigma_2\cdot B_{EVZ}\qquad(3.57)$$

Consideration of compatibility in tan-   Betrachtung der tangentialen Ver-
gential displacement **v**:   schiebungskompatibilität **v**:

$$\frac{2\sigma_2\cdot a}{E_1}\left(1-\mu_1^2\right)\sin2\varphi-\frac{2}{3}\frac{p_2\cdot a}{E_1}\left(2-\mu_1\right)\sin2\varphi+2p_2a\frac{\mu_1^2}{E_1}\sin2\varphi+\frac{q\cdot a}{3E_1}\left(\mu_1-5\right)\sin2\varphi$$

$$+2qa\frac{\mu_1^2}{E_1}\sin2\varphi=\frac{2}{3}\frac{p_2\cdot a}{E_2}\mu_2\sin2\varphi+\frac{2}{3}p_2a\frac{\mu_2^2}{E_2}\sin2\varphi+\frac{q\cdot a}{3E_2}\left(3+\mu_2\right)\sin2\varphi-\frac{2}{3}qa\frac{\mu_2^2}{E_2}\sin2\varphi$$

$$(3.58)$$

$$\sigma_2\frac{6E_2\left(1-\mu_1^2\right)}{3E_1E_2}+p_2\frac{2E_2\left(\mu_1-2+3\mu_1^2\right)-2E_1\mu_2\left(1+\mu_2\right)}{3E_1E_2}=q\frac{E_1\left(3+\mu_2-2\mu_2^2\right)-E_2\left(\mu_1-5+6\mu_1^2\right)}{3E_1E_2}$$

$$p_2=q\underbrace{\frac{E_1\left(3+\mu_2-2\mu_2^2\right)-E_2\left(\mu_1-5+6\mu_1^2\right)}{2E_2\left(\mu_1-2+3\mu_1^2\right)-2E_1\mu_2\left(1+\mu_2\right)}}_{C_{EVZ}}+\sigma_2\underbrace{\frac{6E_2\left(-1+\mu_1^2\right)}{2E_2\left(\mu_1-2+3\mu_1^2\right)-2E_1\mu_2\left(1+\mu_2\right)}}_{D_{EVZ}}$$

$$(3.59)$$

With the abbreviations   Mit den Abkürzungen

$$C_{EVZ}=\frac{E_1\left(3+\mu_2-2\mu_2^2\right)-E_2\left(\mu_1-5+6\mu_1^2\right)}{2E_2\left(\mu_1-2+3\mu_1^2\right)-2E_1\mu_2\left(1+\mu_2\right)}\quad\text{und}\quad D_{EVZ}=\frac{6E_2\left(-1+\mu_1^2\right)}{2E_2\left(\mu_1-2+3\mu_1^2\right)-2E_1\mu_2\left(1+\mu_2\right)}$$

$$(3.60)$$

follows: folgt:

$$p_2 = q \cdot C_{EVZ} + \sigma_2 \cdot D_{EVZ} \tag{3.61}$$

With Equations(3.57) and (3.61) q and p_2 are determined:

Mit (3.57) und (3.61) lässt sich nun q und p_2 bestimmen:

$$q \cdot A_{EVZ} - \sigma_2 \cdot B_{EVZ} = q \cdot C_{EVZ} + \sigma_2 \cdot D_{EVZ}$$

$$\boxed{q = \sigma_2 \frac{B_{EVZ} + D_{EVZ}}{A_{EVZ} - C_{EVZ}}} \tag{3.62}$$

$$p_2 = \sigma_2 \frac{B_{EVZ} + D_{EVZ}}{A_{EVZ} - C_{EVZ}} \cdot A_{EVZ} - \sigma_2 \cdot B_{EVZ}$$

$$p_2 = \sigma_2 \frac{A_{EVZ}\left(B_{EVZ} + D_{EVZ}\right) - B_{EVZ}\left(A_{EVZ} - C_{EVZ}\right)}{A_{EVZ} - C_{EVZ}}$$

$$\boxed{p_2 = \sigma_2 \frac{A_{EVZ}D_{EVZ} + B_{EVZ}C_{EVZ}}{A_{EVZ} - C_{EVZ}}} \tag{3.63}$$

Consideration of compatibility in radial displacement **u**:

Betrachtung der radialen Verschiebungskompatibilität **u**:

$$\frac{\sigma_2 \cdot a}{E_1}\left(1 - \mu_1^2\right)\left(1 - 2\cos 2\varphi\right) - \frac{a}{E_1}\left[p_0\left(1 + \mu_1\right) + \frac{p_2}{3}\left(\mu_1 - 5\right)\cos 2\varphi\right] - 2p_2 a \frac{\mu_1^2}{E_1}\cos 2\varphi$$

$$+\frac{2}{3}\frac{a \cdot q}{E_1}\left(2 - \mu_1\right)\cos 2\varphi - 2qa\frac{\mu_1^2}{E_1}\cos 2\varphi = \frac{a}{E_2}\left[p_0\left(1 - \mu_2\right) - \frac{p_2}{3}\left(3 + \mu_2\right)\cos 2\varphi\right] - 2p_0 a\frac{\mu_2^2}{E_2}$$

$$+\frac{2}{3}p_2 a\frac{\mu_2^2}{E_2}\cos 2\varphi - \frac{2q \cdot a}{3E_2}\mu_2\cos 2\varphi - \frac{2}{3}qa\frac{\mu_2^2}{E_2}\cos 2\varphi$$

$$\tag{3.64}$$

$$p_0 \underbrace{\frac{3E_1\left(1 - \mu_2 - 2\mu_2^2\right) + 3E_2\left(1 + \mu_1\right)}{3E_2\left(1 - \mu_1^2\right)}}_{E_{EVZ}} = \sigma_2\left(1 - 2\cos 2\varphi\right)$$

$$+p_2 \underbrace{\frac{E_1\left(3 + \mu_2 - 2\mu_2^2\right) + E_2\left(5 - \mu_1 - 6\mu_1^2\right)}{3E_2\left(1 - \mu_1^2\right)}}_{H_{EVZ}}\cos 2\varphi + q\underbrace{\frac{2E_1\mu_2\left(1 + \mu_2\right) + 2E_2\left(2 - \mu_1 - 3\mu_1^2\right)}{3E_2\left(1 - \mu_1^2\right)}}_{G_{EVZ}}\cos 2\varphi$$

It should be noted at this point that the term with the stress load σ_2 for practical reasons is not formulated as a fraction term compared to the Volume 1 [67].

An dieser Stelle wird darauf hingewiesen, dass der Term mit der Spannungslast σ_2 aus praktischen Gründen nennerfrei gegenüber dem 1. Band [67] ist.

With the abbreviations

Mit den Abkürzungen

$$E_{EVZ} = \frac{3E_1\left(1 - \mu_2 - 2\mu_2^2\right) + 3E_2\left(1 + \mu_1\right)}{3E_2\left(1 - \mu_1^2\right)}$$

$$G_{EVZ} = \frac{2E_1\mu_2\left(1 + \mu_2\right) + 2E_2\left(2 - \mu_1 - 3\mu_1^2\right)}{3E_2\left(1 - \mu_1^2\right)} \qquad (3.65)$$

$$H_{EVZ} = \frac{E_1\left(3 + \mu_2 - 2\mu_2^2\right) + E_2\left(5 - \mu_1 - 6\mu_1^2\right)}{3E_2\left(1 - \mu_1^2\right)}$$

follows: folgt:

$$p_0 = \frac{1}{E_{EVZ}}\left[\sigma_2\left(1 - 2\cos 2\varphi\right) + p_2 H_{EVZ}\cos 2\varphi + q G_{EVZ}\cos 2\varphi\right]$$

$$p_0 = \frac{1}{E_{EVZ}}\left[\sigma_2 + \underbrace{\left(-2\sigma_2 + p_2 H_{EVZ} + q G_{EVZ}\right)}_{=0}\cos 2\varphi\right] \qquad (3.66)$$

With Equations (3.62) and (3.63) ther-   Unter Berücksichtigung von (3.62) und
fore   (3.63) ergibt sich für

$$\left(-2\sigma_2 + p_2 H_{EVZ} + q G_{EVZ}\right) = 0$$

and transforming gives: und es verbleibt:

$$\boxed{p_0 = \frac{\sigma_2}{E_{EVZ}}} \qquad (3.67)$$

This determines the three unknown   Damit sind die drei unbekannten Pa-
parameters (q, p_0, p_2) of section force   rameter (q, p_0, p_2) der Schnittkraft-
functions X_r, X_φ relative to the circu-   funktionen X_r, X_φ bezogen auf die
lar disc.   Kreisscheibe bestimmt.

$$\boxed{\begin{aligned} X_r(\varphi) &= \frac{\sigma_2}{E_{EVZ}} - \sigma_2\frac{A_{EVZ}D_{EVZ} + B_{EVZ}C_{EVZ}}{A_{EVZ} - C_{EVZ}}\cos 2\varphi \\ X_\varphi(\varphi) &= \sigma_2\frac{B_{EVZ} + D_{EVZ}}{A_{EVZ} - C_{EVZ}}\sin 2\varphi \end{aligned}} \qquad (3.68)$$

They have the exclusive dependence on   Sie besitzen die ausschließliche Abhän-
the material sizes $\left(E_1, E_2, \mu_1, \mu_2\right)$. The   gigkeit von den Materialgrößen
geometric size „Radius" (a) is not a $\left(E_1, E_2, \mu_1, \mu_2\right)$. Die geometrische Größe
dependent parameter. This corresponds   „Radius" (a) stellt keinen abhängigen
entirely to the expected result for the   Parameter dar. Dies entspricht zur
infinitely extended disc.   Gänze dem erwarteten Ergebnis bei der
                                           unendlich ausgedehnten Scheibe.

3.1.9 Mechanical quantities for the disc with circular inclusion

With the well-known section force functions, the mechanical quantities can be formulated separately by means of superposition for the disc

3.1.9 Mechanische Größen der Scheibe mit kreisrundem Einschluss

Mit den nun bekannten Schnittkraftfunktionen können die mechanischen Größen mittels Superposition für die Scheibe

$$S_S = S_{S,0} + S_{S,X_r} + S_{S,X_\varphi} \tag{3.69}$$

and the inclusion:

und den Kern

$$S_K = S_{K,X_r} + S_{K,X_\varphi} \tag{3.70}$$

Attention must be paid to the correct direction according to superimposing compatibility, see section 3.1.8.

getrennt formuliert werden. Bei der Superposition ist auf die richtungstreue Überlagerung entsprechend der Kompatibilität, siehe Abschnitt 3.1.8, zu achten.

3.1.9.1 Stresses

The **stresses** are obtained by the superposition of the stresses on the cut-free structures (disc and inclusion). The quantities of the disc and of the inclusion are to be considered separately.

3.1.9.1 Spannungen

Die **Spannungen** werden durch die Superposition der Spannungen an den freigeschnittenen Strukturen (Scheibe und Kern) gewonnen. Dabei sind die Größen der Scheibe und des Kernes getrennt zu betrachten.

Disc:

Radial stress:

Scheibe:

Radialspannung:

$$\sigma_{S,r,EVZ} = \frac{\sigma_2}{2}\left[1 - \frac{a^2}{r^2} - \left(1 - 4\frac{a^2}{r^2} + 3\frac{a^4}{r^4}\right)\cos 2\varphi\right]$$
$$+ p_0\frac{a^2}{r^2} - p_2\left(2\frac{a^2}{r^2} - \frac{a^4}{r^4}\right)\cos 2\varphi$$
$$- 2q\left(\frac{a^2}{r^2} - \frac{a^4}{r^4}\right)\cos 2\varphi \tag{3.71}$$

Tangential stress:

Tangentialspannung:

$$\sigma_{S,\varphi,EVZ} = \frac{\sigma_2}{2}\left[1+\frac{a^2}{r^2}+\left(1+3\frac{a^4}{r^4}\right)\cos 2\varphi\right]$$

$$-p_0\frac{a^2}{r^2}-p_2\frac{a^4}{r^4}\cos 2\varphi \qquad (3.72)$$

$$-2q\frac{a^4}{r^4}\cos 2\varphi$$

Shear stress: Schubspannung:

$$\tau_{S,r\varphi,EVZ} = \left[\frac{\sigma_2}{2}\left(1+2\frac{a^2}{r^2}-3\frac{a^4}{r^4}\right)-p_2\left(\frac{a^2}{r^2}-\frac{a^4}{r^4}\right)-2q\left(\frac{a^2}{2r^2}-\frac{a^4}{r^4}\right)\right]\sin 2\varphi \qquad (3.73)$$

Inclusion: **Kern:**

Radial stress: Radialspannung:

$$\sigma_{K,r,EVZ} = p_0 - p_2\cos 2\varphi \qquad (3.74)$$

Tangential stress: Tangentialspannung:

$$\sigma_{K,\varphi,EVZ} = p_0 - \left(-p_2+2p_2\frac{r^2}{a^2}\right)\cos 2\varphi + 2q\frac{r^2}{a^2}\cos 2\varphi$$

$$= p_0 + \left[p_2\left(1-2\frac{r^2}{a^2}\right)+2q\frac{r^2}{a^2}\right]\cos 2\varphi \qquad (3.75)$$

Shear stress: Schubspannung:

$$\tau_{K,r\varphi,EVZ} = \left(p_2 - p_2\frac{r^2}{a^2}\right)\sin 2\varphi + q\frac{r^2}{a^2}\sin 2\varphi$$

$$= \left[p_2\left(1-\frac{r^2}{a^2}\right)+q\frac{r^2}{a^2}\right]\sin 2\varphi \qquad (3.76)$$

3.1.9.2 Strains

3.1.9.2 Verzerrungen

The **strains** can be obtained from the stresses by the constitutive relation (3.6) or by superposition of the strains (Sections 3.1.3 to 3.1.7):

Die **Verzerrungen** können aus den Spannungen mithilfe der konstitutiven Beziehung (3.6) oder mittels Superposition der Verzerrungen (Abschnitt 3.1.3 bis 3.1.7) gewonnen werden:

Disc: **Scheibe:**

Radial strain: Radialdehnung:

$$\varepsilon_{S,r,EVZ} = \frac{1}{E_1}\left\{\frac{\sigma_2}{2}\left[1 - \mu_1 - \frac{a^2}{r^2}(1+\mu_1) - 2\mu_1^2 - \left(1 + 3\frac{a^4}{r^4} - 4\frac{a^2}{r^2}(1-\mu_1)\right)(1+\mu_1)\cos 2\varphi\right]\right.$$
$$+ p_0\frac{a^2}{r^2}(1+\mu_1) - p_2\left(2\frac{a^2}{r^2} - \frac{a^4}{r^4}(1+\mu_1) - 2\mu_1^2\frac{a^2}{r^2}\right)\cos 2\varphi$$
$$\left.- 2q\left(\frac{a^2}{r^2} - \frac{a^4}{r^4}(1+\mu_1) - \mu_1^2\frac{a^2}{r^2}\right)\cos 2\varphi\right\}$$

$$(3.77)$$

Tangential strains: Tangentialdehnung:

$$\varepsilon_{S,\varphi,EVZ} = \frac{1}{E_1}\left\{\frac{\sigma_2}{2}\left[1 - \mu_1 + \frac{a^2}{r^2}(1+\mu_1) - 2\mu_1^2 + \left(\left(1 + 3\frac{a^4}{r^4} - 4\mu_1\frac{a^2}{r^2}\right)(1+\mu_1)\right)\cos 2\varphi\right]\right.$$
$$- p_0\frac{a^2}{r^2}(1+\mu_1) + p_2\left(2\mu_1\frac{a^2}{r^2} - \frac{a^4}{r^4}(1+\mu_1) + 2\mu_1^2\frac{a^2}{r^2}\right)\cos 2\varphi \qquad (3.78)$$
$$\left.+ 2q\left(\mu_1\frac{a^2}{r^2} - \frac{a^4}{r^4}(1+\mu_1) + \mu_1^2\frac{a^2}{r^2}\right)\cos 2\varphi\right\}$$

Shear strain: Gleitung:

$$\gamma_{S,r\varphi,EVZ} = \frac{1}{G_1}\left\{\frac{\sigma_2}{2}\left(1 + 2\frac{a^2}{r^2} - 3\frac{a^4}{r^4}\right) - p_2\left(\frac{a^2}{r^2} - \frac{a^4}{r^4}\right) + 2q\left(-\frac{a^2}{2r^2} + \frac{a^4}{r^4}\right)\right\}\sin 2\varphi \quad (3.79)$$

Inclusion: **Kern:**

Radial strain: Radialdehnung:

$$\varepsilon_{K,r,EVZ} = \frac{1}{E_2}\left\{p_0\left(1 - \mu_2 - 2\mu_2^2\right) - p_2\left(1 + \mu_2 - 2\mu_2\frac{r^2}{a^2} - 2\mu_2^2\frac{r^2}{a^2}\right)\cos 2\varphi - 2q\left(\mu_2\frac{r^2}{a^2} + \mu_2^2\frac{r^2}{a^2}\right)\cos 2\varphi\right\}$$

$$(3.80)$$

Tangential strains: Tangentialdehnung:

$$\varepsilon_{K,\varphi,EVZ} = \frac{1}{E_2}\left\{p_0\left(1 - \mu_2 - 2\mu_2^2\right) + p_2\left(1 + \mu_2 - 2\frac{r^2}{a^2} + 2\mu_2^2\frac{r^2}{a^2}\right)\cos 2\varphi + 2q\frac{r^2}{a^2}\left(1 - \mu_2^2\right)\cos 2\varphi\right\}$$

$$(3.81)$$

Shear strain: Gleitung:

$$\gamma_{K,r\varphi,EVZ} = \frac{1}{G_2}\left[p_2\left(1 - \frac{r^2}{a^2}\right) + q\frac{r^2}{a^2}\right]\sin 2\varphi \qquad (3.82)$$

3.1.9.3 Displacements

The **displacements** are obtained also by superposition

3.1.9.3 Verschiebungen

Die **Verschiebungen** werden ebenso durch Superposition erhalten.

Disc:

Scheibe:

Radial displacement:

Radialverschiebung:

$$u_{S,EVZ} = \frac{\sigma_2}{2E_1}\left\{r\left(1-\mu_1\right)+\frac{a^2}{r}\left(1+\mu_1\right)-2r\mu_1^2+\left(-4\frac{a^2}{r}-\left(r-\frac{a^4}{r^3}\right)\left(1+\mu_1\right)+4\mu_1^2\frac{a^2}{r}\right)\cos 2\varphi\right\}$$

$$-\frac{1}{E_1}\left\{p_0\frac{a^2}{r}\left(1+\mu_1\right)+p_2\left[-2\frac{a^2}{r}+\frac{a^4}{3r^3}\left(1+\mu_1\right)+2\mu_1^2\frac{a^2}{r}\right]\cos 2\varphi\right\}$$

$$+\frac{2q}{3E_1}\left[3\frac{a^2}{r}-\left(1+\mu_1\right)\frac{a^4}{r^3}-3\mu_1^2\frac{a^2}{r}\right]\cos 2\varphi$$

$$= u_{S,N_2}$$

$$(3.83)$$

Tangential displacement:

Tangentialverschiebung:

$$v_{S,EVZ} = \frac{\sigma_2}{2E_1}\left[2\frac{a^2}{r}\left(1-\mu_1\right)+\left(r+\frac{a^4}{r^3}\right)\left(1+\mu_1\right)-4\mu_1^2\frac{a^2}{r}\right]\sin 2\varphi$$

$$-\frac{p_2}{E_1}\left[\frac{a^2}{r}\left(1-\mu_1\right)+\frac{1}{3}\frac{a^4}{r^3}\left(1+\mu_1\right)-2\mu_1^2\frac{a^2}{r}\right]\sin 2\varphi$$

$$-\frac{q}{E_1}\left[\left(1-\mu_1\right)\frac{a^2}{r}+\frac{2}{3}\left(1+\mu_1\right)\frac{a^4}{r^3}-2\mu_1^2\frac{a^2}{r}\right]\sin 2\varphi$$

$$= v_{S,N_2}$$

$$(3.84)$$

Inclusion:

Kern:

Radial displacement:

Radialverschiebung:

$$u_{K,EVZ} = \frac{1}{E_2}\left\{p_0\left(1-\mu_2-2\mu_2^2\right)r-p_2\left[\left(1+\mu_2\right)r-\frac{2}{3}\mu_2\frac{r^3}{a^2}-\frac{2}{3}\mu_2^2\frac{r^3}{a^2}\right]\cos 2\varphi-\frac{2q}{3}\mu_2\left(1+\mu_2\right)\frac{r^3}{a^2}\cos 2\varphi\right\}$$

$$= u_{K,N_2}$$

$$(3.85)$$

Tangential displacement:

Tangentialverschiebung:

$$v_{K,EVZ} = \frac{1}{E_2}\left\{p_2\left[r\left(1+\mu_2\right)-\frac{1}{3}\frac{r^3}{a^2}\left(3+\mu_2\right)+\frac{2}{3}\mu_2^2\frac{r^3}{a^2}\right]-\frac{1}{3}q\left(-3-\mu_2+2\mu_2^2\right)\frac{r^3}{a^2}\right\}\sin 2\varphi$$

$$= v_{K,N_2}$$

$$(3.86)$$

3.1.10 Disc with hole under unidirectional force in x-direction

The solution for the disc with circular inclusion under the uniaxial load in the x-direction can be derived in the same form as for the uniaxial load in the y-direction, see section 3.1.1. It is much easier to obtain the solution by the transformation of the coordinate system (rotation with $\alpha = \dfrac{\pi}{2}$).

3.1.10 Scheibe mit Loch unter einachsialer Belastung in x-Richtung

Die Lösung für die Scheibe mit Kern unter der einachsialen Belastung in x-Richtung kann nun in derselben Form wie für die einachsiale Belastung in y-Richtung hergeleitet, siehe Abschnitt 3.1.1, oder wesentlich einfacher durch die Transformation des Koordinatensystems (Drehung um $\alpha = \dfrac{\pi}{2}$) gewonnen werden.

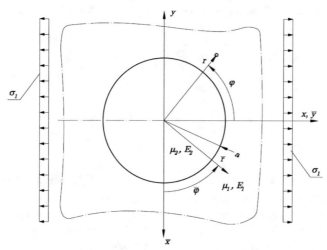

Fig. 3-19: Transformation of coordinate system.
Transformation des Koordinatensystems.

The transformation of the Cartesian $\bar{x}, \bar{y}$-COS into the x, y-COS leads to the following transformation relationship in polar coordinates:

Die Transformation des kartesischen $\bar{x}, \bar{y}$-KOS in das x, y-KOS ergibt die folgende Transformationsbeziehung in Polarkoordinaten:

$$\bar{r} = r$$
$$\bar{\varphi} + \alpha = \bar{\varphi} + \frac{\pi}{2} = \varphi \tag{3.87}$$

The coordinate r remains congruent to $\bar{r}$ and the coordinate φ leads to sign changes in the trigonometric terms:

Die Koordinate r bleibt zu $\bar{r}$ kongruent und die Koordinate φ führt zu einem Vorzeichenwechsel bei den trigonometrischen Termen:

$$\sin\left[2\left(\bar{\varphi} + \frac{\pi}{2}\right)\right] = -\sin 2\varphi$$

$$\cos\left[2\left(\bar{\varphi} + \frac{\pi}{2}\right)\right] = -\cos 2\varphi$$

(3.88)

The three unknown parameters of the section force functions for the stress load σ_1 in the x-direction are described by

Die drei unbekannten Parameter der Schnittkraftfunktionen lauten für die Spannungslast σ_1 in x-Richtung

$$q = \sigma_1 \frac{B_{EVZ} + D_{EVZ}}{A_{EVZ} - C_{EVZ}}$$

$$p_0 = \frac{\sigma_1}{E_{EVZ}}$$

$$p_2 = \sigma_1 \frac{A_{EVZ} D_{EVZ} + B_{EVZ} C_{EVZ}}{A_{EVZ} - C_{EVZ}}$$

(3.89)

The abbreviations A, B, C, D and E remain identical to (3.56), (3.60) and (3.65). The following equations result for the mechanical quantities.

wobei die Abkürzungen A, B, C, D und E identisch zu (3.56), (3.60) und (3.65) bleiben. Für die mechanischen Größen ergeben sich somit die folgenden Gleichungen.

3.1.10.1 Stresses

Disc:

Radial stress:

3.1.10.1 Spannungen

Scheibe:

Radialspannung:

$$\sigma_{S,r,EVZ} = \frac{\sigma_1}{2}\left[1 - \frac{a^2}{r^2} + \left(1 - 4\frac{a^2}{r^2} + 3\frac{a^4}{r^4}\right)\cos 2\varphi\right]$$

$$+ p_0 \frac{a^2}{r^2} + p_2\left(2\frac{a^2}{r^2} - \frac{a^4}{r^4}\right)\cos 2\varphi$$

$$+ 2q\left(\frac{a^2}{r^2} - \frac{a^4}{r^4}\right)\cos 2\varphi$$

(3.90)

Tangential stress: Tangentialspannung:

$$\sigma_{S,\varphi,EVZ} = \frac{\sigma_1}{2}\left[1 + \frac{a^2}{r^2} - \left(1 + 3\frac{a^4}{r^4}\right)\cos 2\varphi\right]$$
$$- p_0 \frac{a^2}{r^2} + p_2 \frac{a^4}{r^4}\cos 2\varphi \qquad (3.91)$$
$$+ 2q\frac{a^4}{r^4}\cos 2\varphi$$

Shear stress: Schubspannung

$$\tau_{S,r\varphi,EVZ} = \left[-\frac{\sigma_1}{2}\left(1 + 2\frac{a^2}{r^2} - 3\frac{a^4}{r^4}\right) + p_2\left(\frac{a^2}{r^2} - \frac{a^4}{r^4}\right) + 2q\left(\frac{a^2}{2r^2} - \frac{a^4}{r^4}\right)\right]\sin 2\varphi \qquad (3.92)$$

Inclusion: **Kern:**

Radial stress: Radialspannung:

$$\sigma_{K,r,EVZ} = p_0 + p_2 \cos 2\varphi \qquad (3.93)$$

Tangential stress: Tangentialspannung:

$$\sigma_{K,\varphi,EVZ} = p_0 + \left[p_2\left(-1 + 2\frac{r^2}{a^2}\right) - 2q\frac{r^2}{a^2}\right]\cos 2\varphi \qquad (3.94)$$

Shear stress: Schubspannung:

$$\tau_{K,r\varphi,EVZ} = \left[p_2\left(-1 + \frac{r^2}{a^2}\right) - q\frac{r^2}{a^2}\right]\sin 2\varphi \qquad (3.95)$$

3.1.10.2 Strains   ## 3.1.10.2 Verzerrungen

Disc: **Scheibe:**

Radial strain: Radialdehnung:

$$\varepsilon_{S,r,EVZ} = \frac{1}{E_1} \left\{ \frac{\sigma_1}{2} \left[1 - \mu_1 - \frac{a^2}{r^2}(1 + \mu_1) - 2\mu_1^2 + \left(1 + 3\frac{a^4}{r^4} - 4\frac{a^2}{r^2}(1 - \mu_1) \right)(1 + \mu_1)\cos 2\varphi \right] \right.$$
$$+ p_0 \frac{a^2}{r^2}(1 + \mu_1) + p_2 \left(2\frac{a^2}{r^2} - \frac{a^4}{r^4}(1 + \mu_1) - 2\mu_1^2 \frac{a^2}{r^2} \right)\cos 2\varphi$$
$$\left. + 2q\left(\frac{a^2}{r^2} - \frac{a^4}{r^4}(1 + \mu_1) - \mu_1^2 \frac{a^2}{r^2} \right)\cos 2\varphi \right\}$$

$$(3.96)$$

Tangential strain: Tangentialdehnung:

$$\varepsilon_{S,\varphi,EVZ} = \frac{1}{E_1} \left\{ \frac{\sigma_1}{2} \left[1 - \mu_1 + \frac{a^2}{r^2}(1 + \mu_1) - 2\mu_1^2 - \left(\left(1 + 3\frac{a^4}{r^4} - 4\mu_1 \frac{a^2}{r^2} \right)(1 + \mu_1) \right)\cos 2\varphi \right] \right.$$
$$- p_0 \frac{a^2}{r^2}(1 + \mu_1) - p_2 \left(2\mu_1 \frac{a^2}{r^2} - \frac{a^4}{r^4}(1 + \mu_1) + 2\mu_1^2 \frac{a^2}{r^2} \right)\cos 2\varphi$$
$$\left. - 2q\left(\mu_1 \frac{a^2}{r^2} - \frac{a^4}{r^4}(1 + \mu_1) + \mu_1^2 \frac{a^2}{r^2} \right)\cos 2\varphi \right\}$$

$$(3.97)$$

Shear strain: Gleitung:

$$\gamma_{S,r\varphi,EVZ} = -\frac{1}{G_1} \left\{ \frac{\sigma_1}{2} \left(1 + 2\frac{a^2}{r^2} - 3\frac{a^4}{r^4} \right) - p_2 \left(\frac{a^2}{r^2} - \frac{a^4}{r^4} \right) + 2q\left(-\frac{a^2}{2r^2} + \frac{a^4}{r^4} \right) \right\}\sin 2\varphi \quad (3.98)$$

Inclusion: **Kern:**

Radial strain: Radialdehnung:

$$\varepsilon_{K,r,EVZ} = \frac{1}{E_2} \left\{ p_0\left(1 - \mu_2 - 2\mu_2^2 \right) + p_2 \left(1 + \mu_2 - 2\mu_2 \frac{r^2}{a^2} - 2\mu_2^2 \frac{r^2}{a^2} \right)\cos 2\varphi + 2q\left(\mu_2 \frac{r^2}{a^2} + \mu_2^2 \frac{r^2}{a^2} \right)\cos 2\varphi \right\}$$

$$(3.99)$$

Tangential strain: Tangentialdehnung:

$$\varepsilon_{K,\varphi,EVZ} = \frac{1}{E_2} \left\{ p_0\left(1 - \mu_2 - 2\mu_2^2 \right) - p_2 \left(1 + \mu_2 - 2\frac{r^2}{a^2} + 2\mu_2^2 \frac{r^2}{a^2} \right)\cos 2\varphi - 2q \frac{r^2}{a^2}\left(1 - \mu_2^2 \right)\cos 2\varphi \right\}$$

$$(3.100)$$

Shear strain: Gleitung:

$$\gamma_{K,r\varphi,EVZ} = -\frac{1}{G_2} \left[p_2\left(1 - \frac{r^2}{a^2} \right) + q\frac{r^2}{a^2} \right]\sin 2\varphi$$

$$(3.101)$$

3.1.10.3 Displacements

Disc:

Radial displacement:

3.1.10.3 Verschiebungen

Scheibe:

Radialverschiebung:

$$
\begin{aligned}
u_{S,EVZ} = \frac{\sigma_2}{2E_1} &\left\{ r(1-\mu_1) + \frac{a^2}{r}(1+\mu_1) - 2r\mu_1^2 - \left(-4\frac{a^2}{r} - \left(r - \frac{a^4}{r^3}\right)(1+\mu_1) + 4\mu_1^2\frac{a^2}{r}\right)\cos 2\varphi \right\} \\
&- \frac{1}{E_1}\left\{ p_0\frac{a^2}{r}(1+\mu_1) - p_2\left[-2\frac{a^2}{r} + \frac{a^4}{3r^3}(1+\mu_1) + 2\mu_1^2\frac{a^2}{r}\right]\cos 2\varphi \right\} \\
&- \frac{2q}{3E_1}\left[3\frac{a^2}{r} - (1+\mu_1)\frac{a^4}{r^3} - 3\mu_1^2\frac{a^2}{r}\right]\cos 2\varphi \\
&= u_{S,N_1}
\end{aligned}
$$

(3.102)

Tangential displacement:

Tangentialverschiebung:

$$
\begin{aligned}
v_{S,EVZ} = -\frac{\sigma_2}{2E_1} &\left[2\frac{a^2}{r}(1-\mu_1) + \left(r + \frac{a^4}{r^3}\right)(1+\mu_1) - 4\mu_1^2\frac{a^2}{r}\right]\sin 2\varphi \\
&+ \frac{p_2}{E_1}\left[\frac{a^2}{r}(1-\mu_1) + \frac{1}{3}\frac{a^4}{r^3}(1+\mu_1) - 2\mu_1^2\frac{a^2}{r}\right]\sin 2\varphi \\
&+ \frac{q}{E_1}\left[(1-\mu_1)\frac{a^2}{r} + \frac{2}{3}(1+\mu_1)\frac{a^4}{r^3} - 2\mu_1^2\frac{a^2}{r}\right]\sin 2\varphi \\
&= v_{S,N_1}
\end{aligned}
$$

(3.103)

Inclusion:

Radial displacement:

Kern:

Radialverschiebung:

$$
\begin{aligned}
u_{K,EVZ} = \frac{1}{E_2} &\left\{ p_0(1-\mu_2-2\mu_2^2)r + p_2\left[(1+\mu_2)r - \frac{2}{3}\mu_2\frac{r^3}{a^2} - \frac{2}{3}\mu_2^2\frac{r^3}{a^2}\right]\cos 2\varphi + \frac{2q}{3}\mu_2(1+\mu_2)\frac{r^3}{a^2}\cos 2\varphi \right\} \\
&= u_{K,N_1}
\end{aligned}
$$

(3.104)

Tangential displacement:

Tangentialverschiebung:

$$
\begin{aligned}
v_{K,EVZ} = -\frac{1}{E_2} &\left\{ p_2\left[r(1+\mu_2) - \frac{1}{3}\frac{r^3}{a^2}(3+\mu_2) + \frac{2}{3}\mu_2^2\frac{r^3}{a^2}\right] - \frac{1}{3}q\left(-3-\mu_2+2\mu_2^2\right)\frac{r^3}{a^2}\right\}\sin 2\varphi \\
&= v_{K,N_1}
\end{aligned}
$$

(3.105)

3.2 Shear force

This section describes for the disc with circular inclusion under shear force the derivation of stresses, strains and deformations. Following the same principle as in section 3.1, the mechanical quantities are formulated at the free edge of the hole for the disc with hole under shear load. Likewise, the mechanical quantities are derived on the circular disc. With the displacement functions and the statically indeterminate section force functions the compatibility can be set up afterwards. From this, the unknown parameters of the statically indeterminate section force functions can be obtained.

3.2 Schubbelastung

In diesem Abschnitt wird die Herleitung der Spannungen, Verzerrungen und Verformungen für die Scheibe mit Kern unter Schubbelastung beschrieben. Nach demselben Prinzip wie im Abschnitt 3.1 werden am freien Lochrand die mechanischen Größen für die Scheibe mit Loch, jedoch nun unter Schubbelastung, formuliert. Ebenso werden die mechanischen Größen an der Kreisscheibe hergeleitet. Mit den Verschiebungsfunktionen und den statisch unbestimmten Schnittkraftfunktionen kann anschließend die Kompatibilität aufgestellt werden. Daraus lassen sich die unbekannten Parameter der statisch unbestimmten Schnittkraftfunktionen gewinnen.

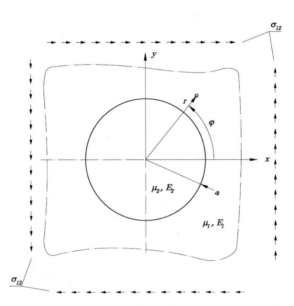

Fig. 3-20: Disc (E_1, μ_1) with inclusion (E_2, μ_2) under shear force (σ_{12}).
Scheibe (E_1, μ_1) mit Kern (E_2, μ_2) unter Schubbelastung (σ_{12}).

3.2.1 Procedure

The formulations take place with the polar coordinates r and φ. In the first step the structure of the disc with circular inclusion is cut by an imaginary section at the circular contact line of inclusion and disc. As unknown section force functions, the functions

3.2.1 Vorgehensweise

Die Formulierungen erfolgen mit den Polarkoordinaten r und φ. Der Kern wird im 1. Schritt von der Scheibe entlang des Berührkreises freigeschnitten. Als unbekannte Schnittkraftfunktionen werden die Funktionen

$$X_r = p_2 \sin 2\varphi$$
$$X_\varphi = q \cos 2\varphi \qquad (3.106)$$

are applied at the circular contact line of inclusion and disc.

In the next step, the displacement functions are formulated for the cut structural parts, disc with hole and circular disc. Their unknown parameters (p_2, q) are determined again using the displacement compatibility provided by a linear system of equations.

am freigeschnittenen Berührkreis angesetzt.

Im nächsten Schritt werden die Verschiebungsfunktionen für die freigeschnittenen Strukturteile, Scheibe mit Loch und Kreisscheibe, formuliert. Ihre unbekannten Parameter (p_2, q) werden wieder mithilfe der Verschiebungskompatibilität, welche ein lineares Gleichungssystem liefert, bestimmt.

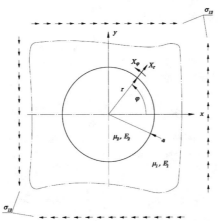

Fig. 3-21: Section-forces at the disc with circular inclusion under shear force.
Schnittkräfte an der Scheibe mit Kern unter Schubbelastung.

3.2.2 Disc with hole under shear load

In analogous form as in Section 3.1.3, the solution for the EVZ can be found by extension the solution for the ESZ [67].

The stress functions for the disc with hole under shear force are:

3.2.2 Scheibe mit Loch unter Schubbelastung

In analoger Form wie im Abschnitt 3.1.3 kann ausgehend von der Lösung für den ESZ [67] durch Erweiterung die Lösung für den EVZ gefunden werden.

Die Spannungsfunktionen der Scheibe mit Loch unter Schubbelastung lauten:

$$\sigma_r = \sigma_{12}\left(1 - 4\frac{a^2}{r^2} + 3\frac{a^4}{r^4}\right)\sin 2\varphi$$

$$\sigma_\varphi = -\sigma_{12}\left(1 + 3\frac{a^4}{r^4}\right)\sin 2\varphi$$

$$\sigma_z = -4\sigma_{12}\mu\frac{a^2}{r^2}\sin 2\varphi \tag{3.107}$$

$$\tau_{r\varphi} = \sigma_{12}\left(1 + 2\frac{a^2}{r^2} - 3\frac{a^4}{r^4}\right)\cos 2\varphi$$

The strains can be formulated with reference to the ESZ:

Die Verzerrungen lassen sich mit dem Bezug zum ESZ formulieren:

$$\varepsilon_{r,EVZ} = \varepsilon_{r,ESZ} - \frac{\mu}{E}\sigma_z$$

$$= \frac{\sigma_{12}}{E}\left[1 + \mu - 4\frac{a^2}{r^2}\left(1 - \mu^2\right) + 3\frac{a^4}{r^4}\left(1 + \mu\right)\right]\sin 2\varphi$$

$$\varepsilon_{\varphi,EVZ} = \varepsilon_{\varphi,ESZ} - \frac{\mu}{E}\sigma_z$$

$$= \frac{\sigma_{12}}{E}\left[-1 - \mu + 4\mu\frac{a^2}{r^2}\left(1 + \mu\right) - 3\frac{a^4}{r^4}\left(1 + \mu\right)\right]\sin 2\varphi \tag{3.108}$$

$$\gamma_{r\varphi,EVZ} = \gamma_{r\varphi,ESZ}$$

$$= \frac{\sigma_{12}}{G}\left(1 + 2\frac{a^2}{r^2} - 3\frac{a^4}{r^4}\right)\cos 2\varphi$$

The displacements at the edge of the hole can be also formulated with reference to the ESZ:

Die Verschiebungen am Lochleibungsrand können ebenso mit dem Bezug zum ESZ formuliert werden:

$$u_{EVZ} = \int\varepsilon_{r,ESZ}\mathrm{d}r + \int 4\frac{a^2}{r^2}\frac{\mu^2}{E}\sigma_{12}\sin 2\varphi\mathrm{d}r$$

$$= u_{ESZ} - 4\frac{a^2}{r}\frac{\mu^2}{E}\sigma_{12}\sin 2\varphi \tag{3.109}$$

$$V_{EVZ} = \int \left(r\varepsilon_{\varphi,EVZ} - u_{EVZ} \right) \mathrm{d}\varphi$$

$$= V_{ESZ} - 4\frac{a^2}{r}\frac{\mu^2}{E}\sigma_{12}\cos 2\varphi \tag{3.110}$$

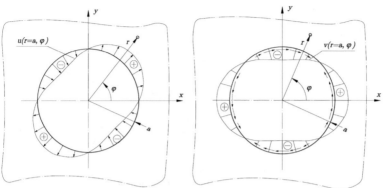

Fig. 3-22: Radial- and tangential deformation at the edge of the hole at the disc with hole under shear force. Radial- und Tangentialverformung am Lochleibungsrand der Scheibe mit Loch unter Schubbelastung.

3.2.3 Disc with hole under radial bearing load

For the infinite disc with hole and the following radial stress load, this represents a trigonometric function

3.2.3 Lochscheibe unter radialer Lochleibungslast

Für die unendliche Scheibe mit Loch mit der trigonometrischen radialen Spannungslast

$$p(\varphi) = p_2 \sin 2\varphi \;, \tag{3.111}$$

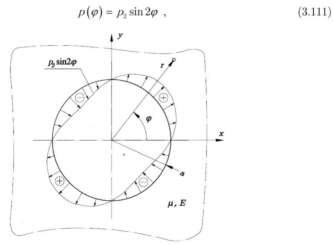

Fig. 3-23: Radial bearing load on the disc with hole.
Radiale Lochleibungslast an der Scheibe mit Loch.

the associated stress components in the plane (r, φ) for plane strain state (EVZ) are:

am Lochleibungsrand lauten die Spannungskomponenten im EVZ:

$$\sigma_r = (-2\frac{a^2}{r^2} + \frac{a^4}{r^4})p_2 \sin 2\varphi$$

$$\sigma_\varphi = -\frac{a^4}{r^4}p_2 \sin 2\varphi$$

$$\sigma_z = -2\frac{a^2}{r^2}\mu p_2 \sin 2\varphi \tag{3.112}$$

$$\tau_{r\varphi} = (\frac{a^2}{r^2} - \frac{a^4}{r^4})p_2 \cos 2\varphi$$

From the associated strains

Aus den zugehörigen Verzerrungen

$$\varepsilon_{r,EVZ} = \varepsilon_{r,ESZ} + 2p_2\frac{\mu^2}{E}\frac{a^2}{r^2}\sin 2\varphi$$

$$= \frac{p_2}{E}\left[-2\frac{a^2}{r^2}\left(1-\mu^2\right) + \frac{a^4}{r^4}\left(1+\mu\right)\right]\sin 2\varphi$$

$$\varepsilon_{\varphi,EVZ} = \varepsilon_{\varphi,ESZ} + 2p_2\frac{\mu^2}{E}\frac{a^2}{r^2}\sin 2\varphi$$

$$= \frac{p_2}{E}\left[2\frac{a^2}{r^2}\left(\mu+\mu^2\right) - \frac{a^4}{r^4}\left(1+\mu\right)\right]\sin 2\varphi \tag{3.113}$$

$$\gamma_{r\varphi,EVZ} = \gamma_{r\varphi,ESZ}$$

$$= \frac{p_2}{G}\left(\frac{a^2}{r^2} - \frac{a^4}{r^4}\right)\cos 2\varphi$$

the displacements can be formulated:

lassen sich die Verschiebungen formulieren:

$$u_{EVZ} = u_{ESZ} - 2p_2\frac{a^2}{r}\frac{\mu^2}{E}\sin 2\varphi \tag{3.114}$$

$$v_{EVZ} = v_{ESZ} - 2p_2\frac{a^2}{r}\frac{\mu^2}{E}\cos 2\varphi \tag{3.115}$$

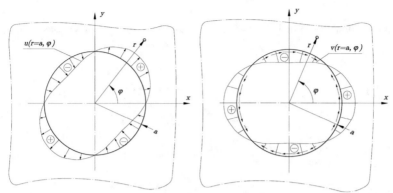

Fig. 3-24: Radial and tangential deformation at the hole edge on the disc with
hole under radial bearing load. Radial- und Tangentialverformung
am Lochleibungsrand unter radialer Lochleibungslast.

3.2.4 Disc with hole under tangential bearing load

3.2.4 Scheibe mit Loch unter tangentialer Lochleibungslast

For the infinite disc with hole and the following trigonometric tangential stress load

Für die unendliche Scheibe mit Loch mit der trigonometrischen tangentialen Spannungslast

$$q(\varphi) = q \cos 2\varphi \ , \tag{3.116}$$

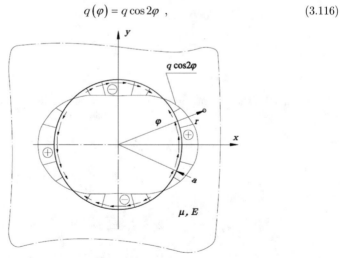

Fig. 3-25: Tangential bearing load on the disc with hole.
Tangentiale Lochleibungslast an der Scheibe mit Loch.

the stress components at EVZ are: ergeben sich die Spannungskomponen-
                                                     ten im EVZ zu:

$$\sigma_r = 2(-\frac{a^2}{r^2} + \frac{a^4}{r^4})q\sin 2\varphi$$

$$\sigma_\varphi = -2\frac{a^4}{r^4}q\sin 2\varphi$$

$$\sigma_z = -2\frac{a^2}{r^2}\mu q\sin 2\varphi \tag{3.117}$$

$$\tau_{r\varphi} = (\frac{a^2}{r^2} - 2\frac{a^4}{r^4})q\cos 2\varphi$$

The strains are: Die Verzerrungen lauten:

$$\varepsilon_{r,EVZ} = \varepsilon_{r,ESZ} + 2q\frac{\mu^2}{E}\frac{a^2}{r^2}\sin 2\varphi$$

$$= \frac{2q}{E}\left[-\frac{a^2}{r^2}\left(1 - \mu^2\right) + \frac{a^4}{r^4}(1 + \mu)\right]\sin 2\varphi$$

$$\varepsilon_{\varphi,EVZ} = \varepsilon_{\varphi,ESZ} + 2q\frac{\mu^2}{E}\frac{a^2}{r^2}\sin 2\varphi \tag{3.118}$$

$$= \frac{2q}{E}\left[\frac{a^2}{r^2}\left(\mu + \mu^2\right) - \frac{a^4}{r^4}(1 + \mu)\right]\sin 2\varphi$$

$$\gamma_{r\varphi,EVZ} = \gamma_{r\varphi,ESZ}$$

$$= \frac{q}{G}\left(\frac{a^2}{r^2} - 2\frac{a^4}{r^4}\right)\cos 2\varphi$$

The associated displacements result in: Die zugehörigen Verschiebungen erge-
                                                     ben sich zu:

$$u_{EVZ} = u_{ESZ} - 2q\frac{a^2}{r}\frac{\mu^2}{E}\sin 2\varphi$$

$$= 2\frac{q}{E}\left[\frac{a^2}{r}\left(1 - \mu^2\right) - \frac{a^4}{3r^3}(1 + \mu)\right]\sin 2\varphi \tag{3.119}$$

$$v_{EVZ} = v_{ESZ} - 2q\frac{a^2}{r}\frac{\mu^2}{E}\cos 2\varphi$$

$$= \frac{q}{E}\left[\frac{a^2}{r}\left(1 - \mu - 2\mu^2\right) + \frac{2}{3}\frac{a^4}{r^3}(1 + \mu)\right]\cos 2\varphi \tag{3.120}$$

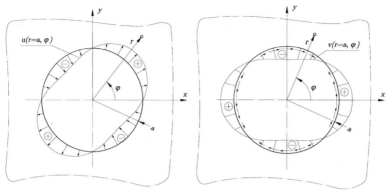

Fig. 3-26: Radial and tangential deformation at the hole edge on the disc with hole under radial bearing load. Radial- und Tangentialverformung am Lochleibungsrand unter tangentialer Lochleibungslast.

3.2.5 Circular disc under radial edge load

The circular disc is loaded at the outer edge by the radial trigonometric stress load:

3.2.5 Kreisscheibe unter radialer Randlast

Die Kreisscheibe wird am Außenrand mit der folgenden radialen trigonometrischen Spannungslast belastet:

$$p(\varphi) = p_2 \sin 2\varphi \qquad (3.121)$$

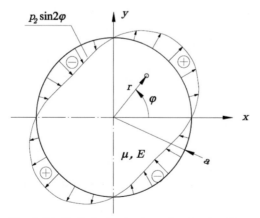

Fig. 3-27: Radial edge load on the circular disc. Radiale Randlast an der Kreisscheibe.

The associated stress components at EVZ are:

Die zugehörigen Spannungskomponenten im EVZ lauten:

$$\sigma_r = p_2 \sin 2\varphi$$

$$\sigma_\varphi = p_2(-1 + 2\frac{r^2}{a^2}) \sin 2\varphi$$

$$\sigma_z = 2\frac{r^2}{a^2} \mu p_2 \sin 2\varphi \tag{3.122}$$

$$\tau_{r\varphi} = p_2(1 - \frac{r^2}{a^2}) \cos 2\varphi$$

With the strains

Mit den Verzerrungen

$$\varepsilon_{r,EVZ} = \varepsilon_{r,ESZ} - 2p_2 \frac{\mu^2}{E} \frac{r^2}{a^2} \sin 2\varphi$$

$$= \frac{1}{E}\left[1 + \mu - 2\frac{r^2}{a^2}\mu(1+\mu)\right]p_2 \sin 2\varphi$$

$$\varepsilon_{\varphi,EVZ} = \varepsilon_{\varphi,ESZ} - 2p_2 \frac{\mu^2}{E} \frac{r^2}{a^2} \sin 2\varphi \tag{3.123}$$

$$= \frac{1}{E}\left[-1 - \mu + 2\frac{r^2}{a^2}(1-\mu^2)\right]p_2 \sin 2\varphi$$

$$\gamma_{r\varphi,EVZ} = \gamma_{r\varphi,ESZ}$$

$$= \frac{p_2}{G}\left(1 - \frac{r^2}{a^2}\right)\cos 2\varphi$$

the displacements results in:

ergeben sich die Verschiebungen:

$$u_{EVZ} = u_{ESZ} - \frac{2}{3}p_2 \frac{r^3}{a^2}\frac{\mu^2}{E}\sin 2\varphi$$

$$= \frac{p_2}{E}\left[r(1+\mu) - \frac{2}{3}\frac{r^3}{a^2}\mu(1+\mu)\right]\sin 2\varphi \tag{3.124}$$

$$v_{EVZ} = v_{ESZ} + \frac{2}{3}p_2 \frac{r^3}{a^2}\frac{\mu^2}{E}\cos 2\varphi$$

$$= \frac{p_2}{E}\left[r(1+\mu) - \frac{1}{3}\frac{r^3}{a^2}(3+\mu-2\mu)\right]\cos 2\varphi \tag{3.125}$$

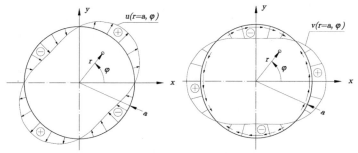

Fig. 3-28: Radial and tangential deformation at the outer edge on the circular disc under radial edge load. Radial- und Tangentialverformung am Kreisscheibenrand unter radialer Randlast.

3.2.6 Circular disc under tangential edge load

The circular disc is loaded at the outer edge by the tangential stress load:

3.2.6 Kreisscheibe unter tangentialer Randlast

Die Kreisscheibe wird am Rand durch die tangentiale Spannungslast

$$q(\varphi) = q\cos 2\varphi \tag{3.126}$$

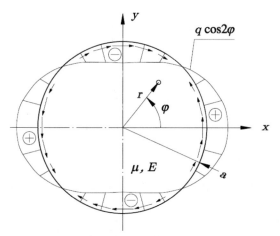

Fig. 3-29: Tangential edge load on the circular disc.
Tangentiale Randlast an der Kreisscheibe.

The associated stress components at EVZ are:

Die zugehörigen Spannungskomponenten im EVZ lauten:

$$\sigma_r = 0$$

$$\sigma_\varphi = -2q\frac{r^2}{a^2}\sin 2\varphi$$

$$\sigma_z = -2q\frac{r^2}{a^2}\mu\sin 2\varphi \tag{3.127}$$

$$\tau_{r\varphi} = q\frac{r^2}{a^2}\cos 2\varphi$$

The constitutive relationship (3.6) provides the strains:

Mit der konstitutiven Beziehung (3.6) folgt für die Verzerrungen:

$$\varepsilon_{r,EVZ} = \varepsilon_{r,ESZ} + 2q\frac{\mu^2}{E}\frac{r^2}{a^2}\sin 2\varphi$$

$$= \frac{2q}{E}\frac{r^2}{a^2}\left[\mu + \mu^2\right]\sin 2\varphi$$

$$\varepsilon_{\varphi,EVZ} = \varepsilon_{\varphi,ESZ} + 2q\frac{\mu^2}{E}\frac{r^2}{a^2}\sin 2\varphi \tag{3.128}$$

$$= \frac{2q}{E}\frac{r^2}{a^2}\left[-1 + \mu^2\right]\sin 2\varphi$$

$$\gamma_{r\varphi} = \frac{q}{G}\frac{r^2}{a^2}\cos 2\varphi$$

By integrating the radial displacement component

Durch Integration wird die radiale Verschiebungskomponente

$$u_{EVZ} = u_{ESZ} + \frac{2}{3}q\frac{\mu^2}{E}\frac{r^3}{a^2}\sin 2\varphi$$

$$= \frac{2}{3}\frac{q}{E}\frac{r^3}{a^2}\left(\mu + \mu^2\right)\sin 2\varphi \tag{3.129}$$

and the tangential displacement component is found:

und tangentiale Verschiebungskomponente gefunden:

$$v_{EVZ} = v_{ESZ} + \frac{q}{3E}\frac{r^3}{a^2}\left(3 + \mu\right)\cos 2\varphi$$

$$= \frac{q}{3E}\frac{r^3}{a^2}\left(3 + \mu - 2\mu^2\right)\cos 2\varphi \tag{3.130}$$

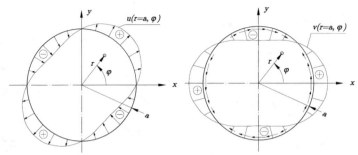

Fig. 3-30: Radial and tangential deformation at the outer edge on the circular
disc under tangential edge load. Radial- und Tangentialverformung
am Kreisscheibenrand unter tangentialer Randlast.

Hence all the required edge deformations are known and applying compatibility the statically indeterminates
(p_2, q) can be determined.

Damit sind nun alle erforderlichen
Randverformungen bekannt und es
kann wiederum die Kompatibilität zur
Bestimmung der statisch unbestimmten Größen (p_2, q) durchgeführt werden.

3.2.7 Compatibility

Analogous to section 3.1.8, the displacement compatibility can now be
used with the known displacement
functions at the cut boundary between
the disc and the inclusion. This is
based on the deformation of the disc
with hole, which is now reduced with
the statically indeterminate parameters. The resulting reduced displacement in the disc with hole corresponds
to the displacement of the circular disc
due to the statically indeterminate
quantities. It should be noted that the
displacements are applied in correct
direction. The following illustrations
show these displacements in the correct
direction on the cut system.

3.2.7 Kompatibilität

Analog zu Abschnitt 3.1.8 kann jetzt
mit den bekannten Verschiebungsfunktionen an der freigeschnittenen Berandung zwischen Scheibe und Kern die
Verschiebungskompatibilität   angewandt werden. Dazu wird wieder von
der Verformung an der Scheibe mit
Loch ausgegangen, welche nun mit den
statisch unbestimmten Größen reduziert wird. Diese sich dadurch reduzierte Verschiebung in der Scheibe entspricht der Verschiebung der Kernscheibe zufolge der statisch unbestimmten Größen. Der richtungstreue
Ansatz der Verschiebungen wird in den
nachfolgenden Abbildungen gezeigt.

Disc with hole under shear load σ_{12}:   **Scheibe mit Loch unter Schublast σ_{12}:**

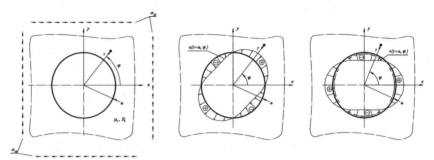

Fig. 3-31: Shear load and deformations $u(r = a,\varphi)$, $v(r = a,\varphi)$, at the edge of the hole on the disc with hole. Schublast σ_{12} und Verformungen $u(r = a,\varphi)$, $v(r = a,\varphi)$ am Lochrand an der Scheibe mit Loch.

Radial deformation: Radialverschiebung:

$$u_{EVZ}(a) = 4a\sigma_{12}\frac{\left(1-\mu_1^2\right)}{E_1}\sin 2\varphi = u_0 \tag{3.131}$$

Tangential deformation: Tangentialverschiebung:

$$v_{EVZ}(a) = 4a\sigma_{12}\frac{\left(1-\mu_1^2\right)}{E_1}\cos 2\varphi = v_0 \tag{3.132}$$

Disc with hole under radial bearing load $X_r(\varphi) = -p_2\sin 2\varphi$:   **Scheibe mit Loch unter radialer Lochleibungslast $X_r(\varphi) = -p_2\sin 2\varphi$:**

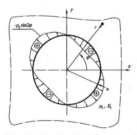
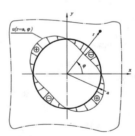
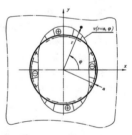

Fig. 3-32: Radial bearing load $X_r(\varphi) = -p_2\sin 2\varphi$ and deformations $u(r = a,\varphi)$, $v(r = a,\varphi)$, at the edge of the hole on the disc with hole. Radiale Lochleibungslast $X_r(\varphi) = -p_2\sin 2\varphi$ und Verformungen $u(r = a,\varphi)$, $v(r = a,\varphi)$ am Lochrand an der Scheibe mit Loch.

Radial deformation: Radialverschiebung:

$$u_{EVZ}(a) = \frac{a}{3E_1} p_2 \left(-5 + \mu_1 + 6\mu_1^2\right)\sin 2\varphi = u_{S,X_r} \tag{3.133}$$

Tangential deformation: Tangentialverschiebung:

$$v_{EVZ}(a) = \frac{2a}{3E_1} p_2 \left(-2 + \mu_1 + 3\mu_1^2\right)\cos 2\varphi = v_{S,X_r} \tag{3.134}$$

Disc with hole under tangential bearing load $X_\varphi(\varphi) = -q\cos 2\varphi$:   **Scheibe mit Loch unter tangentialer Lochleibungslast** $X_\varphi(\varphi) = -q\cos 2\varphi$:

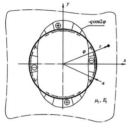 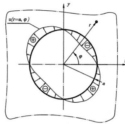 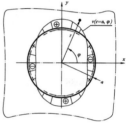

Fig. 3-33: Tangential bearing load $X_\varphi(\varphi) = -q\cos 2\varphi$ and deformations
$u(r = a,\varphi)$, $v(r = a,\varphi)$ at the edge of the hole on the disc with hole.
Tangentiale Lochleibungslast $X_\varphi(\varphi) = -q\cos 2\varphi$ und Verformungen
$u(r = a,\varphi)$, $v(r = a,\varphi)$ am Lochrand an der Scheibe mit Loch.

Radial deformation :   Radialverschiebung:

$$u_{EVZ}(a) = \frac{2a}{3E_1} q\left(-2 + \mu_1 + 3\mu_1^2\right)\sin 2\varphi = u_{S,X_\varphi} \tag{3.135}$$

Tangential deformation: Tangentialverschiebung

$$v_{EVZ}(a) = \frac{a}{3E_1} q\left(-5 + \mu_1 + 6\mu_1^2\right)\cos 2\varphi = v_{S,X_\varphi} \tag{3.136}$$

Circular disc under radial bearing load
$X_r(\varphi) = p_2 \sin 2\varphi$:

Kreisscheibe unter radialer Randlast
$X_r(\varphi) = p_2 \sin 2\varphi$:

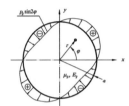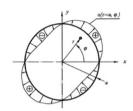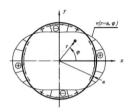

Fig. 3-34: Radial edge load $X_r(\varphi) = p_2 \sin 2\varphi$ and deformations $u(r = a, \varphi)$, $v(r = a, \varphi)$ at the circular disc edge. Radiale Randlast $X_r(\varphi) = p_2 \sin 2\varphi$ und Verformungen $u(r = a, \varphi)$, $v(r = a, \varphi)$ am Kreisscheibenrand.

Radial deformation :   Radialverschiebung:

$$u_{EVZ}(a) = \frac{a}{3E_2} p_2 \left(3 + \mu_2 - 2\mu_2^2\right)\sin 2\varphi = u_{K,X_r} \qquad (3.137)$$

Tangential deformation: Tangentialverschiebung

$$v_{EVZ}(a) = \frac{2a}{3E_2} p_2 \mu_2 \left(1 + \mu_2\right)\cos 2\varphi = v_{K,X_r} \qquad (3.138)$$

Circular disc under tangential bearing load $X_\varphi(\varphi) = q \cos 2\varphi$:

Kreisscheibe unter tangentialer Randlast $X_\varphi(\varphi) = q \cos 2\varphi$:

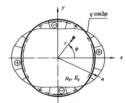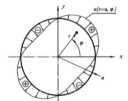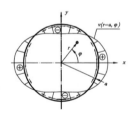

Fig. 3-35: Tangential edge load $X_\varphi(\varphi) = q \cos 2\varphi$ and deformations $u(r = a, \varphi)$, $v(r = a, \varphi)$ at the circular disc edge. Tangentiale Randlast $X_\varphi(\varphi) = q \cos 2\varphi$ und Verformungen $u(r = a, \varphi)$, $v(r = a, \varphi)$ am Kreisscheibenrand.

Radial deformation :   Radialverschiebung:

$$u_{EVZ}(a) = \frac{2a}{3E_2} q \mu_2 \left(1 + \mu_2\right)\sin 2\varphi = u_{K,X_\varphi} \qquad (3.139)$$

Tangential deformation: Tangentialverschiebung:

$$v_{EVZ}\left(a\right) = \frac{a}{3E_2} q\left(3 + \mu_2 - 2\mu_2^2\right)\cos 2\varphi = v_{K,X_\varphi}$$ (3.140)

Compatibility of displacement: **Verschiebungskompatibilität:**

By equating the displacements in the   Durch Gleichsetzen der Verschiebungen
disc and in the inclusion, the displace-   in der Scheibe und im Kern ergibt sich
ment compatibility results: die Verschiebungskompatibilität:

$$u_0 + \left(u_{S,X_r} + u_{S,X_\varphi}\right) = u_{K,X_r} + u_{K,X_\varphi}$$

$$v_0 + \left(v_{S,X_r} + v_{S,X_\varphi}\right) = v_{K,X_r} + v_{K,X_\varphi}$$ (3.141)

It provides two equations for the two   Sie liefert zwei Gleichungen für die
unknown parameters $\left(q, p_2\right)$ of the   zwei unbekannten Parameter $\left(q, p_2\right)$
section force functions X_r, X_φ. With   der Schnittkraftfunktionen X_r, X_φ. Mit
(3.131) to (3.140) the equations (3.141)   (3.131) bis (3.140) gehen die Gleichun-
become: gen (3.141) über in

u:

$$\frac{4a}{E_1}\sigma_{12}\left(1 - \mu_1^2\right)\sin 2\varphi + \left[\frac{a}{3E_1} p_2\left(-5 + \mu_1 + 6\mu_1^2\right)\sin 2\varphi + \frac{2a}{3E_1} q\left(-2 + \mu_1 + 3\mu_1^2\right)\sin 2\varphi\right] =$$

$$\frac{a}{3E_2} p_2\left(3 + \mu_2 - 2\mu_2^2\right)\sin 2\varphi + \frac{2a}{3E_2} q\mu_2\left(1 + \mu_2\right)\sin 2\varphi$$

v:

$$\frac{4a}{E_1}\sigma_{12}\left(1 - \mu_1^2\right)\cos 2\varphi + \left[\frac{2a}{3E_1} p_2\left(-2 + \mu_1 + 3\mu_1^2\right)\cos 2\varphi + \frac{a}{3E_1} q\left(-5 + \mu_1 + 6\mu_1^2\right)\cos 2\varphi\right] =$$

$$\frac{2a}{3E_2} p_2\mu_2\left(1 + \mu_2\right)\cos 2\varphi + \frac{a}{3E_2} q\left(3 + \mu_2 - 2\mu_2^2\right)\cos 2\varphi$$

(3.142)

They can be simplified: die sich vereinfachen lassen zu

$$\frac{E_1\left(3 + \mu_2 - 2\mu_2^2\right) - E_2\left(-5 + \mu_1 + 6\mu_1^2\right)}{3E_1E_2} p_2 = \frac{12\left(1 - \mu_1^2\right)E_2}{3E_1E_2}\sigma_{12} + 2\frac{-E_1\mu_2\left(1 + \mu_2\right) + E_2\left(-2 + \mu_1 + 3\mu_1^2\right)}{3E_1E_2} q$$

$$2\frac{E_1\mu_2\left(1 + \mu_2\right) + E_2\left(-2 + \mu_1 + \mu_1^2\right)}{3E_1E_2} p_2 = \frac{12\left(1 - \mu_1^2\right)E_2}{3E_1E_2}\sigma_{12} + \frac{-E_1\left(3 + \mu_2 - 2\mu_2^2\right) + E_2\left(-5 + \mu_1 + 6\mu_1^2\right)}{3E_1E_2} q$$

<table>
<tr><td>With the abbreviations</td><td>und mit den Abkürzungen</td></tr>
</table>

$$A_{EVZ} = \frac{E_1\left(3+\mu_2-2\mu_2^2\right)-E_2\left(-5+\mu_1+6\mu_1^2\right)}{3E_1E_2}$$

$$B_{EVZ} = 2\frac{-E_1\mu_2\left(1+\mu_2\right)+E_2\left(-2+\mu_1+3\mu_1^2\right)}{3E_1E_2} \qquad (3.143)$$

$$C_{EVZ} = \frac{12\left(1-\mu_1^2\right)E_2}{3E_1E_2}$$

follows: übergehen in die Form:

$$A_{EVZ}p_2 = B_{EVZ}q + C_{EVZ}\sigma_{12}$$
$$B_{EVZ}p_2 = A_{EVZ}q - C_{EVZ}\sigma_{12}$$

For the parameters, this results in: Für die Parameter ergeben sich somit:

$$\boxed{p_2 = \frac{C_{EVZ}}{A_{EVZ}-B_{EVZ}}\sigma_{12} = q \quad \Rightarrow \quad p_2 = \frac{12E_2\left(1-\mu_1^2\right)}{3E_1\left(1+\mu_2\right)+E_2\left(1+\mu_1\right)}\sigma_{12} = q} \qquad (3.144)$$

Thus, the parameters $\left(q,p_2\right)$ that are considered equivalent to the section force functions X_r, X_φ are determined

Damit sind die Parameter $\left(q,p_2\right)$, die sich als gleichwertig darstellen, der Schnittkraftfunktionen X_r, X_φ bestimmt

$$\boxed{\begin{aligned}X_r\left(\varphi\right) &= \frac{12E_2\left(1-\mu_1^2\right)}{3E_1\left(1+\mu_2\right)+E_2\left(1+\mu_1\right)}\sigma_{12}\sin 2\varphi \\ X_\varphi\left(\varphi\right) &= \frac{12E_2\left(1-\mu_1^2\right)}{3E_1\left(1+\mu_2\right)+E_2\left(1+\mu_1\right)}\sigma_{12}\cos 2\varphi\end{aligned}} \qquad (3.145)$$

and the analytical solutions to the mechanical quantities can be formulated. They have also the exclusive dependence on the material sizes $\left(E_1,E_2,\mu_1,\mu_2\right)$. The geometric size „Radius" $\left(a\right)$ is not a dependent parameter. This corresponds entirely to the expected result for the infinitely extended disc.

und es können die analytischen Lösungen zu den mechanischen Größen formuliert werden. Auch sie besitzen die ausschließliche Abhängigkeit von den Materialgrößen $\left(E_1,E_2,\mu_1,\mu_2\right)$. Die geometrische Größe „Radius" $\left(a\right)$ stellt keinen abhängigen Parameter dar. Dies entspricht wiederum zur Gänze dem erwarteten Ergebnis bei der unendlich ausgedehnten Scheibe.

3.2.8 Mechanical quantities for the disc with circular inclusion

The mechanical quantities can be formulated separately by means of superposition for the disc

3.2.8 Mechanische Größen für die Scheibe mit Kern

Die mechanischen Größen werden wieder mittels Superposition für die Scheibe

$$S_S = S_{S,0} + S_{S,X_r} + S_{S,X_\varphi} \tag{3.146}$$

and the inclusion:

und für den Kern:

$$S_K = S_{K,X_r} + S_{K,X_\varphi} \tag{3.147}$$

Superposition in the correct direction according to compatibility is given in section 3.2.7.

getrennt formuliert. Die richtungstreue Überlagerung entsprechend der Kompatibilität ist dem Abschnitt 3.2.7 zu entnehmen.

3.2.8.1 Stresses

The stresses are obtained by the superposition of the stresses on the cut-free structures (disc and inclusion). The quantities of the disc and of the inclusion are to be considered separately.

3.2.8.1 Spannungen

Die Spannungen werden durch die Superposition der Spannungen an den freigeschnittenen Strukturen (Scheibe und Kern) gewonnen. Dabei sind die Größen der Scheibe und des Kernes getrennt zu betrachten.

Disc:

Radial stress:

Scheibe:

Radialspannung:

$$\sigma_{S,r,EVZ} = \left[\sigma_{12}\left(1 - 4\frac{a^2}{r^2} + 3\frac{a^4}{r^4}\right) + p_2\left(2\frac{a^2}{r^2} - \frac{a^4}{r^4}\right) + 2q\left(\frac{a^2}{r^2} - \frac{a^4}{r^4}\right) \right]\sin 2\varphi \tag{3.148}$$

Tangential stress:

Tangentialspannung:

$$\sigma_{S,\varphi,EVZ} = \left[-\sigma_{12}\left(1 + 3\frac{a^4}{r^4}\right) + p_2\frac{a^4}{r^4} + 2q\frac{a^4}{r^4} \right]\sin 2\varphi \tag{3.149}$$

Shear stress

Schubspannung:

$$\tau_{S,r\varphi,EVZ} = \left[\sigma_{12}\left(1 + 2\frac{a^2}{r^2} - 3\frac{a^4}{r^4}\right) - p_2\left(\frac{a^2}{r^2} - \frac{a^4}{r^4}\right) - q\left(\frac{a^2}{r^2} - 2\frac{a^4}{r^4}\right) \right]\cos 2\varphi \tag{3.150}$$

Inclusion:

Radial stress:

Kern:

Radialspannung:

$$\sigma_{K,r,EVZ} = p_2 \sin 2\varphi \tag{3.151}$$

Tangential stress:

Tangentialspannung:

$$\sigma_{K,\varphi,EVZ} = \left[p_2 \left(-1 + 2\frac{r^2}{a^2} \right) - 2q\frac{r^2}{a^2} \right] \sin 2\varphi \tag{3.152}$$

Shear stress:

Schubspannung:

$$\tau_{K,r\varphi,EVZ} = \left[p_2 \left(1 - \frac{r^2}{a^2} \right) + q\frac{r^2}{a^2} \right] \cos 2\varphi \tag{3.153}$$

3.2.8.2 Strains

The strains can be obtained from the stresses by the constitutive relation (3.6):

3.2.8.2 Verzerrungen

Die Verzerrungen können aus den Spannungen mithilfe der konstitutiven Beziehung (3.6) gewonnen werden:

Disc:

Radial strain:

Scheibe:

Radialdehnung:

$$\varepsilon_{S,r,EVZ} = \varepsilon_{S,0,r,EVZ} + \varepsilon_{S,X_r,r,EVZ} + \varepsilon_{S,X_\varphi,r,EVZ}$$

$$= \frac{\sigma_{12}}{E_1} \left[1 + \mu_1 - 4\frac{a^2}{r^2}\left(1 - \mu_1^2\right) + 3\frac{a^4}{r^4}\left(1 + \mu_1\right) \right] \sin 2\varphi$$

$$- \frac{p_2}{E_1} \left[-2\frac{a^2}{r^2}\left(1 - \mu_1^2\right) + \frac{a^4}{r^4}\left(1 + \mu_1\right) \right] \sin 2\varphi \tag{3.154}$$

$$- \frac{2q}{E_1} \left[-\frac{a^2}{r^2}\left(1 - \mu_1^2\right) + \frac{a^4}{r^4}\left(1 + \mu_1\right) \right] \sin 2\varphi$$

Tangential strain:

Tangentialdehnung:

$$\varepsilon_{S,\varphi,EVZ} = \varepsilon_{S,0,\varphi,EVZ} + \varepsilon_{S,X_r,\varphi,EVZ} + \varepsilon_{S,X_\varphi,\varphi,EVZ}$$

$$= \frac{\sigma_{12}}{E_1} \left[-1 - \mu_1 + 4\mu_1\frac{a^2}{r^2}\left(1 + \mu_1\right) - 3\frac{a^4}{r^4}\left(1 + \mu_1\right) \right] \sin 2\varphi$$

$$- \frac{p_2}{E_1} \left[2\frac{a^2}{r^2}\left(\mu_1 + \mu_1^2\right) - \frac{a^4}{r^4}\left(1 + \mu_1\right) \right] \sin 2\varphi \tag{3.155}$$

$$- \frac{2q}{E_1} \left[\frac{a^2}{r^2}\left(\mu_1 + \mu_1^2\right) - \frac{a^4}{r^4}\left(1 + \mu_1\right) \right] \sin 2\varphi$$

Shear strain: Gleitung:

$$\gamma_{S,r\varphi,EVZ} = \gamma_{S,0,r\varphi,EVZ} + \gamma_{S,X_r,r\varphi,EVZ} + \gamma_{S,X_\varphi,r\varphi,EVZ}$$

$$= \frac{1}{G_1}\left[\sigma_{12}\left(1 + 2\frac{a^2}{r^2} - 3\frac{a^4}{r^4}\right) - p_2\left(\frac{a^2}{r^2} - \frac{a^4}{r^4}\right) - q\left(\frac{a^2}{r^2} - 2\frac{a^4}{r^4}\right)\right]\cos 2\varphi \quad (3.156)$$

Inclusion: **Kern:**

Radial strain: Radialdehnung:

$$\varepsilon_{K,r,EVZ} = \varepsilon_{K,X_r,r,EVZ} + \varepsilon_{K,X_\varphi,r,EVZ}$$

$$= \frac{1}{E_2}\left\{p_2\left[1 + \mu_2 - 2\frac{r^2}{a^2}\mu_2\left(1 + \mu_2\right)\right] + 2q\frac{r^2}{a^2}\left[\mu_2 + \mu_2^2\right]\right\}\sin 2\varphi \quad (3.157)$$

Tangential strain: Tangentialdehnung:

$$\varepsilon_{K,\varphi,EVZ} = \varepsilon_{K,X_r,\varphi,EVZ} + \varepsilon_{K,X_\varphi,\varphi,EVZ}$$

$$= \frac{1}{E_2}\left\{p_2\left[-1 - \mu + 2\frac{r^2}{a^2}\left(1 - \mu^2\right)\right] - 2q\frac{r^2}{a^2}\left(-1 + \mu^2\right)\right\}\sin 2\varphi \quad (3.158)$$

Shear strain: Gleitung

$$\gamma_{K,r\varphi,EVZ} = \frac{1}{G_2}\tau_{K,r\varphi,EVZ}$$

$$= \frac{1}{G_2}\left[p_2\left(1 - \frac{r^2}{a^2}\right) + q\frac{r^2}{a^2}\right]\cos 2\varphi \quad (3.159)$$

3.2.8.3 Displacements   ### 3.2.8.3 Verschiebungen

The displacements are obtained also by   Die Verschiebungen werden ebenso
superposition.   durch Superposition erhalten.

Disc: **Scheibe:**

Radial displacement: Radialverschiebung:

$$u_{S,EVZ} = u_{S,0,EVZ} = u_{S,X_r,EVZ} + u_{S,X_\varphi,EVZ}$$

$$= \frac{\sigma_{12}}{E_1}\left[4\frac{a^2}{r} + \left(r - \frac{a^4}{r^3}\right)\left(1 + \mu_1\right) - 4\frac{a^2}{r}\mu_1^2\right]\sin 2\varphi$$

$$-\frac{p_2}{E_1}\left[2\frac{a^2}{r} - \frac{a^4}{3r^3}\left(1 + \mu_1\right) - 2\frac{a^2}{r}\mu_1^2\right]\sin 2\varphi \quad (3.160)$$

$$-\frac{2q}{E_1}\left[\frac{a^2}{r} - \frac{a^4}{3r^3}\left(1 + \mu_1\right) - \frac{a^2}{r}\mu_1^2\right]\sin 2\varphi$$

$$= u_{S,EVZ,N_{12}}$$

Tangential displacement: Tangentialverschiebung:

$$
\begin{aligned}
v_{S,EVZ} = {} & \frac{\sigma_{12}}{E_1}\left[2\frac{a^2}{r}(1-\mu_1)+\left(r+\frac{a^4}{r^3}\right)(1+\mu_1)-4\frac{a^2}{r}\mu_1^2\right]\cos 2\varphi \\
& -\frac{p_2}{E_1}\left[\frac{a^2}{r}(1-\mu_1)+\frac{1}{3}\frac{a^4}{r^3}(1+\mu_1)-2\frac{a^2}{r}\mu_1^2\right]\cos 2\varphi \\
& -\frac{q}{E_1}\left[\frac{a^2}{r}(1-\mu_1)+\frac{2}{3}\frac{a^4}{r^3}(1+\mu_1)-2\frac{a^2}{r}\mu_1^2\right]\cos 2\varphi \\
& = v_{S,EVZ,N_{12}}
\end{aligned}
\tag{3.161}
$$

Inclusion: **Kern:**

Radial displacement: Radialverschiebung:

$$
u_{K,EVZ} = \frac{p_2}{E_2}\left[r(1+\mu_2)-\frac{2}{3}\frac{r^3}{a^2}\left(\mu_2+\mu_2^2\right)\right]\sin 2\varphi+\frac{2q}{3E_2}\left(\mu_2+\mu_2^2\right)\frac{r^3}{a^2}\sin 2\varphi
\tag{3.162}
$$

$$
= u_{K,EVZ,N_{12}}
$$

Tangential displacement: Tangentialverschiebung:

$$
v_{K,EVZ} = \frac{p_2}{E_2}\left[r(1+\mu_2)-\frac{1}{3}\frac{r^3}{a^2}\left(3+\mu_2-2\mu_2^2\right)\right]\cos 2\varphi+\frac{q}{3E_2}\frac{r^3}{a^2}\left(3+\mu_2-2\mu_2^2\right)\cos 2\varphi
$$

$$
= v_{K,EVZ,N_{12}}
$$

$$
\tag{3.163}
$$

As an alternative and as a check, the stresses can be derived also from the displacement function. For this purpose initially stresses are formulated by means of the constitutive relation (3.6):

Zur Kontrolle können die Spannungen auch alternativ von den Verschiebungen abgeleitet werden. Dazu werden zunächst die Spannungen mithilfe der konstitutiven Beziehung (3.6) formuliert:

$$
\varepsilon_r = \frac{1}{E}\left[\sigma_r-\mu\left(\sigma_\varphi+\sigma_z\right)\right] \rightarrow \sigma_r = E\varepsilon_r+\mu\left(\sigma_\varphi+\sigma_z\right)
$$

$$
\varepsilon_\varphi = \frac{1}{E}\left[\sigma_\varphi-\mu\left(\sigma_r+\sigma_z\right)\right] \rightarrow \sigma_r = \frac{1}{\mu}\left(\sigma_\varphi-E\varepsilon_\varphi-\mu\sigma_z\right)
$$

$$
\varepsilon_z = \frac{1}{E}\left[\sigma_z-\mu\left(\sigma_r+\sigma_\varphi\right)\right] = 0 \rightarrow \sigma_z = \mu\left(\sigma_r+\sigma_\varphi\right)
$$

$$
\mu E\varepsilon_r+\sigma_\varphi\left(1+\mu\right)\mu^2 = \sigma_\varphi\left(1-\mu\right)^2\left(1+\mu\right)-E\varepsilon_\varphi\left(1-\mu\right)
$$

$$\sigma_r = \frac{E}{1+\mu}\left[\varepsilon_r + \frac{\mu}{1-2\mu}\left(\varepsilon_r + \varepsilon_\varphi\right)\right]$$

$$\sigma_\varphi = \frac{E}{1+\mu}\left[\varepsilon_\varphi + \frac{\mu}{1-2\mu}\left(\varepsilon_r + \varepsilon_\varphi\right)\right]$$

(3.164)

$$\gamma = \frac{1}{G}\tau \to \tau = G\gamma$$

(3.165)

With the strains, obtained by the kinematic relationship (3.9)

Mit den Verzerrungen, welche mithilfe der kinematischen Beziehung (3.9)

$$\varepsilon_r = \frac{\partial u}{\partial r}$$

$$\varepsilon_\varphi = \frac{1}{r}\left(u + \frac{\partial v}{\partial\varphi}\right)$$

$$\gamma = \frac{1}{r}\frac{\partial u}{\partial\varphi} + r\frac{\partial}{\partial r}\left(\frac{v}{r}\right)$$

and the well known displacements ((3.160) to (3.163)), the stresses ((3.164) and (3.165)) can be formulated.

aus den bekannten Verschiebungen ((3.160) bis (3.163)) folgen, können die Spannungen (3.164) und (3.165) berechnet werden.

3.3 Plane elasticity – disc with circular inclusion

3.3 Ebenes Elastizitätsgesetz – Scheibe mit Kern

Due to combining the solutions for the unidirectional and shear load, based on the superposition principle, it is possible to build up the plane elasticity law for the disc with circular inclusion. The general constitutive relationship for the EVZ in the Voigt matrix notation is

Durch die Kombination der Lösungen für die Normal- und Schublast, auf Basis des Superpositionsprinzips, lässt sich das ebene Elastizitätsgesetz für die Scheibe mit Kern aufbauen. Die allgemeine konstitutive Beziehung für den EVZ in der Voigt'schen Matrixschreibweise lautet:

$$\boldsymbol{\sigma} = \mathbf{K}\boldsymbol{\varepsilon}$$

$$\boldsymbol{\varepsilon} = \mathbf{J}\boldsymbol{\sigma}$$

(3.166)

$$\begin{pmatrix} \sigma_{11} \\ \sigma_{22} \\ \sigma_{12} \end{pmatrix} = \begin{bmatrix} K_{11} & K_{12} & K_{13} \\ K_{21} & K_{22} & K_{23} \\ K_{31} & K_{32} & K_{33} \end{bmatrix} \begin{pmatrix} \varepsilon_{rr} \\ \varepsilon_{\varphi\varphi} \\ \gamma_{r\varphi} \end{pmatrix}$$

$$\begin{pmatrix} \varepsilon_{rr} \\ \varepsilon_{\varphi\varphi} \\ \gamma_{r\varphi} \end{pmatrix} = \begin{bmatrix} J_{11} & J_{12} & J_{13} \\ J_{21} & J_{22} & J_{23} \\ J_{31} & J_{32} & J_{33} \end{bmatrix} \begin{pmatrix} \sigma_{11} \\ \sigma_{22} \\ \sigma_{12} \end{pmatrix} \qquad (3.167)$$

It should be noted that this notation creates a relationship between cartesian coordinate loads and polar coordinate strains. This reference is needed to combine the solutions accordingly and does not contradict the mechanical understanding. Rather, the relationship can be seen analogously to a standard experiment for determining the material properties of an orthotropic material. In this case, the characteristic material property is determined empirically for each axis or for each plane according to a specified force magnitude. These characteristic material properties are summarized for the three-dimensional material model in a compliance tensor. By inverting the compliance tensor, the stiffness tensor of the orthotropic material is determined, and the constitutive relationship or the law of material is determined completely. The transformation of the strain quantities into Cartesian coordinates is possible without any restrictions after the combination. The stress-describing formulation relies on the stiffness matrix $\mathbf{K}$ and the strain-describing formulation on the compliance matrix $\mathbf{J}$. It is now necessary to describe the elements of these matrices using the known mechanical quantities, which are derived as functions.

Es ist zu beachten, dass mit dieser Schreibweise ein Bezug zwischen Belastungen in kartesischen Koordinaten und Verzerrungen in Polarkoordinaten geschaffen wird. Dieser Bezug ist eben erforderlich, um die Lösungen entsprechend kombinieren zu können und widerspricht nicht dem mechanischen Verständnis. Vielmehr kann der Bezug analog zu einem gewöhnlichen Versuch zur Bestimmung der Materialeigenschaften eines orthotropen Werkstoffes gesehen werden. Dabei wird für jede Achse bzw. für jede Ebene der charakteristische Materialkennwert zufolge einer spezifizierten Kraftgröße empirisch ermittelt. Diese charakteristischen Materialkennwerte werden für das räumliche Materialmodell in einem Nachgiebigkeitstensor zusammengefasst. Durch Invertierung des Nachgiebigkeitstensors wird der Steifigkeitstensor des orthotropen Materials ermittelt und die konstitutive Beziehung bzw. das Materialgesetz ist somit vollständig ermittelt. Eine Transformation der Verzerrungsgrößen in kartesische Koordinaten ist nach der Kombination ohne Einschränkungen möglich. Die spannungsbeschreibende Formulierung stützt sich auf die Steifigkeitsmatrix $\mathbf{K}$ und die dehnungsbeschreibende auf die Nachgiebigkeitsmatrix $\mathbf{J}$.

For this purpose, the strain-describing constitutive relations are formulated individually for the three different loads (uniaxial load in the x-, and y-direction and shear load).

Es gilt nun die Elemente dieser Matrizen $\left(K_{ij}, J_{ij} \right)$ mithilfe der bekannten mechanischen Größen, welche als Funktionen hergeleitet wurden, zu beschreiben. Dazu werden für die drei unterschiedlichen Belastungen (einachsiale Belastung in x-, und y-Richtung und Schubbelastung) die dehnungsbeschreibenden konstitutiven Beziehungen einzeln formuliert.

3.3.1 Compliance – unidirectional load in x-direction

3.3.1 Nachgiebigkeit – Einnachsiale Belastung in x-Richtung

The constitutive relationship for unidirectional loading in the x direction is

Die konstitutive Beziehung bei unidirektionaler Belastung in x-Richtung lautet

$$
\begin{pmatrix} \varepsilon_{rr} \\ \varepsilon_{\varphi\varphi} \\ \gamma_{r\varphi} \end{pmatrix} = \begin{bmatrix} J_{11} & J_{12} & J_{13} \\ J_{21} & J_{22} & J_{23} \\ J_{31} & J_{32} & J_{33} \end{bmatrix} \begin{pmatrix} \sigma_{11} \\ 0 \\ 0 \end{pmatrix}
\tag{3.168}
$$

and the strains are:

woraus sich die Verzerrungen

$$
\varepsilon_{rr} = J_{11}\sigma_{11} \rightarrow J_{11} = \frac{\varepsilon_{rr}}{\sigma_{11}}
$$

$$
\varepsilon_{\varphi\varphi} = J_{21}\sigma_{11} \rightarrow J_{21} = \frac{\varepsilon_{\varphi\varphi}}{\sigma_{11}}
\tag{3.169}
$$

$$
\gamma_{r\varphi} = J_{31}\sigma_{11} \rightarrow J_{31} = \frac{\gamma_{r\varphi}}{\sigma_{11}}
$$

Taking into account the strains, (3.96) to (3.101), the elements J_{i1} of the compliance matrix are given as functions $J_{i1}(r,\varphi)$ of the variables r and φ. Now, the elastic law for plane strain can be formulated separately for the disc and the nucleus.

formulieren lassen. Unter Berücksichtigung der Verzerrungen, (3.96) bis (3.101), ergeben sich die Elemente J_{i1} der Nachgiebigkeitsmatrix als Funktionen $J_{i1}(r,\varphi)$ der Variablen r und φ. Das ebene Elastizitätsgesetz kann nun wiederum für die Scheibe und den Kern getrennt formuliert werden.

Disc: **Scheibe:**

$$J_{S,11,EVZ}(r,\varphi) = \frac{\varepsilon_{S,r,EVZ}(r,\varphi)}{\sigma_1}$$

$$= \frac{1}{E_1}\left\{\frac{1}{2}\left[1 - \mu_1 - \frac{a^2}{r^2}(1+\mu_1) - 2\mu_1^2 + \left(1 + 3\frac{a^4}{r^4} - 4\frac{a^2}{r^2}(1-\mu_1)\right)(1+\mu_1)\cos 2\varphi\right]\right.$$

$$+ \frac{p_{0,\sigma_1}}{\sigma_1}\frac{a^2}{r^2}(1+\mu_1) + \frac{p_{2,\sigma_1}}{\sigma_1}\left(2\frac{a^2}{r^2} - \frac{a^4}{r^4}(1+\mu_1) - 2\mu_1^2\frac{a^2}{r^2}\right)\cos 2\varphi$$

$$\left. + \frac{2q_{\sigma_1}}{\sigma_1}\left(\frac{a^2}{r^2} - \frac{a^4}{r^4}(1+\mu_1) - \mu_1^2\frac{a^2}{r^2}\right)\cos 2\varphi\right\}$$

$$(3.170)$$

$$J_{S,21,EVZ}(r,\varphi) = \frac{\varepsilon_{S,\varphi,EVZ}(r,\varphi)}{\sigma_1}$$

$$= \frac{1}{E_1}\left\{\frac{1}{2}\left[1 - \mu_1 + \frac{a^2}{r^2}(1+\mu_1) - 2\mu_1^2 - \left(\left(1 + 3\frac{a^4}{r^4} - 4\mu_1\frac{a^2}{r^2}\right)(1+\mu_1)\right)\cos 2\varphi\right]\right.$$

$$- \frac{p_{0,\sigma_1}}{\sigma_1}\frac{a^2}{r^2}(1+\mu_1) - \frac{p_{2,\sigma_1}}{\sigma_1}\left(2\mu_1\frac{a^2}{r^2} - \frac{a^4}{r^4}(1+\mu_1) + 2\mu_1^2\frac{a^2}{r^2}\right)\cos 2\varphi$$

$$\left. - \frac{2q_{\sigma_1}}{\sigma_1}\left(\mu_1\frac{a^2}{r^2} - \frac{a^4}{r^4}(1+\mu_1) + \mu_1^2\frac{a^2}{r^2}\right)\cos 2\varphi\right\}$$

$$(3.171)$$

$$J_{S,31,EVZ}(r,\varphi) = \frac{\gamma_{S,r\varphi,EVZ}(r,\varphi)}{\sigma_1}$$

$$= -\frac{1}{G_1}\left\{\frac{1}{2}\left(1 + 2\frac{a^2}{r^2} - 3\frac{a^4}{r^4}\right) - \frac{p_{2,\sigma_1}}{\sigma_1}\left(\frac{a^2}{r^2} - \frac{a^4}{r^4}\right) + \frac{2q_{\sigma_1}}{\sigma_1}\left(-\frac{a^2}{2r^2} + \frac{a^4}{r^4}\right)\right\}\sin 2\varphi$$

$$(3.172)$$

Inclusion: **Kern:**

$$J_{K,11,EVZ}(r,\varphi) = \frac{\varepsilon_{K,r,EVZ}(r,\varphi)}{\sigma_1}$$

$$= \frac{1}{\sigma_1 E_2}\left\{p_{0,\sigma_1}\left(1 - \mu_2 - 2\mu_2^2\right) + p_{2,\sigma_1}\left(1 + \mu_2 - 2\mu_2\frac{r^2}{a^2} - 2\mu_2^2\frac{r^2}{a^2}\right)\cos 2\varphi + 2q_{\sigma_1}\left(\mu_2\frac{r^2}{a^2} + \mu_2^2\frac{r^2}{a^2}\right)\cos 2\varphi\right\}$$

$$(3.173)$$

$$J_{K,21,EVZ}\left(r,\varphi\right)=\frac{\varepsilon_{K,\varphi,EVZ}\left(r,\varphi\right)}{\sigma_1}$$

$$=\frac{1}{\sigma_1 E_2}\left\{p_{0,\sigma_1}\left(1-\mu_2-2\mu_2^2\right)-p_{2,\sigma_1}\left(1+\mu_2-2\frac{r^2}{a^2}+2\mu_2^2\frac{r^2}{a^2}\right)\cos 2\varphi-2q_{\sigma_1}\frac{r^2}{a^2}\left(1-\mu_2^2\right)\cos 2\varphi\right\}$$

$$(3.174)$$

$$J_{K,31,EVZ}\left(r,\varphi\right)=\frac{\gamma_{K,r\varphi,EVZ}\left(r,\varphi\right)}{\sigma_1}$$

$$=-\frac{1}{\sigma_1 G_2}\left[p_{2,\sigma_1}\left(1-\frac{r^2}{a^2}\right)+q_{\sigma_1}\frac{r^2}{a^2}\right]\sin 2\varphi$$

$$(3.175)$$

3.3.2 Compliance – unidirectional load in y-direction

3.3.2 Nachgiebigkeit – Einachsiale Belastung in y-Richtung

With the constitutive relationship

Mit der konstitutiven Beziehung

$$\begin{pmatrix}\varepsilon_{rr}\\\varepsilon_{\varphi\varphi}\\\gamma_{r\varphi}\end{pmatrix}=\begin{bmatrix}J_{11}&J_{12}&J_{13}\\J_{21}&J_{22}&J_{23}\\J_{31}&J_{32}&J_{33}\end{bmatrix}\begin{pmatrix}0\\\sigma_{22}\\0\end{pmatrix}\qquad(3.176)$$

and with unidirectional stress in the y-direction, the strains can be expressed as follows:

lassen sich bei unidirektionaler Belastung in y-Richtung die Verzerrungen wie folgt ausdrücken:

$$\varepsilon_{rr}=J_{12}\sigma_{22}\rightarrow J_{12}=\frac{\varepsilon_{rr}}{\sigma_{22}}$$

$$\varepsilon_{\varphi\varphi}=J_{22}\sigma_{22}\rightarrow J_{22}=\frac{\varepsilon_{\varphi\varphi}}{\sigma_{22}}$$

$$\gamma_{r\varphi}=J_{32}\sigma_{22}\rightarrow J_{32}=\frac{\gamma_{r\varphi}}{\sigma_{22}}$$

$$(3.177)$$

With the strains (3.77) to (3.82) the functions of the compliance matrix $J_{i2}\left(r,\varphi\right)$ are given:

Mit den Verzerrungen (3.77) bis (3.82) ergeben sich die Funktionen $J_{i2}\left(r,\varphi\right)$ der Nachgiebigkeitsmatrix.

Disc: **Scheibe:**

$$J_{S,12,EVZ}\left(r,\varphi\right) = \frac{\varepsilon_{S,r,EVZ}\left(r,\varphi\right)}{\sigma_2}$$

$$= \frac{1}{E_1}\left\{\frac{1}{2}\left[1-\mu_1-\frac{a^2}{r^2}\left(1+\mu_1\right)-2\mu_1^2-\left(1+3\frac{a^4}{r^4}-4\frac{a^2}{r^2}\left(1-\mu_1\right)\right)\left(1+\mu_1\right)\cos 2\varphi\right]\right.$$

$$+\frac{P_{0,\sigma_2}}{\sigma_2}\frac{a^2}{r^2}\left(1+\mu_1\right)-\frac{P_{2,\sigma_2}}{\sigma_2}\left(2\frac{a^2}{r^2}-\frac{a^4}{r^4}\left(1+\mu_1\right)-2\mu_1^2\frac{a^2}{r^2}\right)\cos 2\varphi$$

$$\left.-\frac{2q_{\sigma_2}}{\sigma_2}\left(\frac{a^2}{r^2}-\frac{a^4}{r^4}\left(1+\mu_1\right)-\mu_1^2\frac{a^2}{r^2}\right)\cos 2\varphi\right\}$$

$$(3.178)$$

$$J_{S,22,EVZ}\left(r,\varphi\right) = \frac{\varepsilon_{S,\varphi,EVZ}\left(r,\varphi\right)}{\sigma_2}$$

$$= \frac{1}{E_1}\left\{\frac{1}{2}\left[1-\mu_1+\frac{a^2}{r^2}\left(1+\mu_1\right)-2\mu_1^2+\left(\left(1+3\frac{a^4}{r^4}-4\mu_1\frac{a^2}{r^2}\right)\left(1+\mu_1\right)\right)\cos 2\varphi\right]\right.$$

$$-\frac{P_{0,\sigma_2}}{\sigma_2}\frac{a^2}{r^2}\left(1+\mu_1\right)+\frac{P_{2,\sigma_2}}{\sigma_2}\left(2\mu_1\frac{a^2}{r^2}-\frac{a^4}{r^4}\left(1+\mu_1\right)+2\mu_1^2\frac{a^2}{r^2}\right)\cos 2\varphi$$

$$\left.+\frac{2q_{\sigma_2}}{\sigma_2}\left(\mu_1\frac{a^2}{r^2}-\frac{a^4}{r^4}\left(1+\mu_1\right)+\mu_1^2\frac{a^2}{r^2}\right)\cos 2\varphi\right\}$$

$$(3.179)$$

$$J_{S,32,EVZ}\left(r,\varphi\right) = \frac{\gamma_{S,r\varphi,EVZ}\left(r,\varphi\right)}{\sigma_2}$$

$$= \frac{1}{G_1}\left\{\frac{1}{2}\left(1+2\frac{a^2}{r^2}-3\frac{a^4}{r^4}\right)-\frac{P_{2,\sigma_2}}{\sigma_2}\left(\frac{a^2}{r^2}-\frac{a^4}{r^4}\right)+\frac{2q_{\sigma_2}}{\sigma_2}\left(-\frac{a^2}{2r^2}+\frac{a^4}{r^4}\right)\right\}\sin 2\varphi$$

$$(3.180)$$

Inclusion: **Kern:**

$$J_{K,12,EVZ}\left(r,\varphi\right) = \frac{\varepsilon_{K,r,EVZ}\left(r,\varphi\right)}{\sigma_2}$$

$$= \frac{1}{\sigma_2 E_2}\left\{P_{0,\sigma_2}\left(1-\mu_2-2\mu_2^2\right)-P_{2,\sigma_2}\left(1+\mu_2-2\mu_2\frac{r^2}{a^2}-2\mu_2^2\frac{r^2}{a^2}\right)\cos 2\varphi-2q_{\sigma_2}\left(\mu_2\frac{r^2}{a^2}+\mu_2^2\frac{r^2}{a^2}\right)\cos 2\varphi\right\}$$

$$(3.181)$$

$$J_{K,22,EVZ}(r,\varphi) = \frac{\varepsilon_{K,\varphi,EVZ}(r,\varphi)}{\sigma_2}$$

$$= \frac{1}{\sigma_2 E_2}\left\{p_{0,\sigma_2}\left(1-\mu_2-2\mu_2^2\right)+p_{2,\sigma_2}\left(1+\mu_2-2\frac{r^2}{a^2}+2\mu_2^2\frac{r^2}{a^2}\right)\cos2\varphi+2q_{\sigma_2}\frac{r^2}{a^2}\left(1-\mu_2^2\right)\cos2\varphi\right\}$$

$$(3.182)$$

$$J_{K,32,EVZ}(r,\varphi) = \frac{\gamma_{K,r\varphi,EVZ}(r,\varphi)}{\sigma_2}$$

$$= \frac{1}{\sigma_2 G_2}\left[p_{2,\sigma_2}\left(1-\frac{r^2}{a^2}\right)+q_{\sigma_2}\frac{r^2}{a^2}\right]\sin2\varphi$$

$$(3.183)$$

3.3.3 Compliance – shear load

The constitutive relationship for shear load is:

3.3.3 Nachgiebigkeit – Schubbelastung

Die ebene konstitutive Beziehung bei Schubbelastung lautet:

$$\begin{pmatrix}\varepsilon_{rr}\\\varepsilon_{\varphi\varphi}\\\gamma_{r\varphi}\end{pmatrix} = \begin{bmatrix}J_{11} & J_{12} & J_{13}\\J_{21} & J_{22} & J_{23}\\J_{31} & J_{32} & J_{33}\end{bmatrix}\begin{pmatrix}0\\0\\\sigma_{12}\end{pmatrix}$$

$$(3.184)$$

Hence these strains can be explicitly expressed:

Daraus lassen sich die Verzerrungen explizit ausdrücken:

$$\varepsilon_{rr} = J_{13}\sigma_{12} \rightarrow J_{13} = \frac{\varepsilon_{rr}}{\sigma_{12}}$$

$$\varepsilon_{\varphi\varphi} = J_{23}\sigma_{12} \rightarrow J_{23} = \frac{\varepsilon_{\varphi\varphi}}{\sigma_{12}}$$

$$\gamma_{r\varphi} = J_{33}\sigma_{12} \rightarrow J_{33} = \frac{\gamma_{r\varphi}}{\sigma_{12}}$$

$$(3.185)$$

With the strains (3.154) to (3.159) the functions of the compliance matrix $J_{i3}(r,\varphi)$ can be formulated.

Mit den Verzerrungen (3.154) bis (3.159) werden die Funktionen $J_{i3}(r,\varphi)$ der Nachgiebigkeitsmatrix formuliert.

Disc: **Scheibe:**

$$
\begin{aligned}
J_{S,13,EVZ}\left(r,\varphi\right) &= \frac{\varepsilon_{S,r,EVZ}\left(r,\varphi\right)}{\sigma_{12}} \\
&= \frac{1}{E_1}\left[1+\mu_1-4\frac{a^2}{r^2}\left(1-\mu_1^2\right)+3\frac{a^4}{r^4}\left(1+\mu_1\right)\right]\sin 2\varphi \\
&\quad -\frac{p_{2,\sigma_{12}}}{\sigma_{12}E_1}\left[-2\frac{a^2}{r^2}\left(1-\mu_1^2\right)+\frac{a^4}{r^4}\left(1+\mu_1\right)\right]\sin 2\varphi \\
&\quad -\frac{2q_{\sigma_{12}}}{\sigma_{12}E_1}\left[-\frac{a^2}{r^2}\left(1-\mu_1^2\right)+\frac{a^4}{r^4}\left(1+\mu_1\right)\right]\sin 2\varphi
\end{aligned}
\tag{3.186}
$$

$$
\begin{aligned}
J_{S,23,EVZ}\left(r,\varphi\right) &= \frac{\varepsilon_{S,\varphi,EVZ}\left(r,\varphi\right)}{\sigma_{12}} \\
&= \frac{1}{E_1}\left[-1-\mu_1+4\mu_1\frac{a^2}{r^2}\left(1+\mu_1\right)-3\frac{a^4}{r^4}\left(1+\mu_1\right)\right]\sin 2\varphi \\
&\quad -\frac{p_{2,\sigma_{12}}}{\sigma_{12}E_1}\left[2\frac{a^2}{r^2}\left(\mu_1+\mu_1^2\right)-\frac{a^4}{r^4}\left(1+\mu_1\right)\right]\sin 2\varphi \\
&\quad -\frac{2q_{\sigma_{12}}}{\sigma_{12}E_1}\left[\frac{a^2}{r^2}\left(\mu_1+\mu_1^2\right)-\frac{a^4}{r^4}\left(1+\mu_1\right)\right]\sin 2\varphi
\end{aligned}
\tag{3.187}
$$

$$
\begin{aligned}
J_{S,33,EVZ}\left(r,\varphi\right) &= \frac{\gamma_{S,r\varphi,EVZ}\left(r,\varphi\right)}{\sigma_{12}} \\
&= \frac{1}{G_1}\left[\left(1+2\frac{a^2}{r^2}-3\frac{a^4}{r^4}\right)-\frac{p_{2,\sigma_{12}}}{\sigma_{12}}\left(\frac{a^2}{r^2}-\frac{a^4}{r^4}\right)-\frac{q_{\sigma_{12}}}{\sigma_{12}}\left(\frac{a^2}{r^2}-2\frac{a^4}{r^4}\right)\right]\cos 2\varphi
\end{aligned}
\tag{3.188}
$$

Inclusion: **Kern:**

$$
\begin{aligned}
J_{K,13,EVZ}\left(r,\varphi\right) &= \frac{\varepsilon_{K,r,EVZ}\left(r,\varphi\right)}{\sigma_{12}} \\
&= \frac{1}{\sigma_{12}E_2}\left\{p_{2,\sigma_{12}}\left[1+\mu_2-2\frac{r^2}{a^2}\mu_2\left(1+\mu_2\right)\right]+2q_{\sigma_{12}}\frac{r^2}{a^2}\left[\mu_2+\mu_2^2\right]\right\}\sin 2\varphi
\end{aligned}
\tag{3.189}
$$

$$
\begin{aligned}
J_{K,23,EVZ}\left(r,\varphi\right) &= \frac{\varepsilon_{K,\varphi,EVZ}\left(r,\varphi\right)}{\sigma_{12}} \\
&= \frac{1}{\sigma_{12}E_2}\left\{p_{2,\sigma_{12}}\left[-1-\mu+2\frac{r^2}{a^2}\left(1-\mu^2\right)\right]-2q_{\sigma_{12}}\frac{r^2}{a^2}\left(-1+\mu^2\right)\right\}\sin 2\varphi
\end{aligned}
\tag{3.190}
$$

$$J_{K,33,EVZ}(r,\varphi) = \frac{\gamma_{K,r\varphi,EVZ}(r,\varphi)}{\sigma_{12}}$$

$$= \frac{1}{\sigma_{12}G_2}\left[p_{2,\sigma_{12}}\left(1 - \frac{r^2}{a^2}\right) + q_{\sigma_{12}}\frac{r^2}{a^2}\right]\cos 2\varphi \qquad (3.191)$$

Now all elements of the compliance matrix are formulated and the elastic law for plane strain for the disc with circular inclusion can be formed.

Damit sind nun alle Elemente der Nachgiebigkeitsmatrix formuliert und es kann das ebene Elastizitätsgesetz für die Scheibe mit Kern gebildet werden.

3.3.4 Elastic law for plane strain in matrix notation

3.3.4 Ebenes Elastizitätsgesetz in Matrizenform

The plane elastic law described below is formulated generally and, in particular, is to be used separately for the disc and the inclusion. In general for the disc with circular inclusion, therefore the compliance matrix is occupied with functions.

Das nachfolgend beschriebene ebene Elastizitätsgesetz wird allgemein formuliert und ist im speziellen entsprechend für die Scheibe und den Kern getrennt anzuwenden. Im Allgemeinen ist somit für die Scheibe mit Kern die Nachgiebigkeitsmatrix mit Funktionen besetzt

$$\mathbf{J} = \begin{bmatrix} J_{11,EVZ}(r,\varphi) & J_{12,EVZ}(r,\varphi) & J_{13,EVZ}(r,\varphi) \\ J_{21,EVZ}(r,\varphi) & J_{22,EVZ}(r,\varphi) & J_{23,EVZ}(r,\varphi) \\ J_{31,EVZ}(r,\varphi) & J_{32,EVZ}(r,\varphi) & J_{33,EVZ}(r,\varphi) \end{bmatrix} \qquad (3.192)$$

The compliance matrix can be converted by inversion into the stiffness matrix for the disc with circular inclusion:

und sie kann durch Invertierung in die Steifigkeitsmatrix für die Scheibe mit Kern übergeführt werden:

$$\mathbf{K} = \mathbf{J}^{-1} \qquad (3.193)$$

$$\mathbf{K} = \begin{bmatrix} K_{11,EVZ}(r,\varphi) & K_{12,EVZ}(r,\varphi) & K_{13,EVZ}(r,\varphi) \\ K_{21,EVZ}(r,\varphi) & K_{22,EVZ}(r,\varphi) & K_{23,EVZ}(r,\varphi) \\ K_{31,EVZ}(r,\varphi) & K_{32,EVZ}(r,\varphi) & K_{33,EVZ}(r,\varphi) \end{bmatrix} \qquad (3.194)$$

The matrices are antisymmetric and allow the analytical description of the mechanical quantities for the plane state with any normal and thrust loads.

Die Matrizen sind antimetrisch und erlauben die analytische Beschreibung der mechanischen Größen für das ebene Problem bei beliebigen Normal- und Schublasten.

3.3.4.1 Strains

The strains in polar coordinates can be formulated now for the disc and the inclusion by means of the constitutive relationship.

Disc:

3.3.4.1 Verzerrungen

Die Verzerrungen in Polarkoordinaten lassen sich nun für die Scheibe und den Kern mithilfe der konstitutiven Beziehung formulieren.

Scheibe:

$$\varepsilon_{S,rr,EVZ} = J_{S,11,EVZ}\left(r,\varphi\right)\sigma_{11} + J_{S,12,EVZ}\left(r,\varphi\right)\sigma_{22} + J_{S,13,EVZ}\left(r,\varphi\right)\sigma_{12}$$
$$\varepsilon_{S,\varphi\varphi,EVZ} = J_{S,21,EVZ}\left(r,\varphi\right)\sigma_{11} + J_{S,22,EVZ}\left(r,\varphi\right)\sigma_{22} + J_{S,23,EVZ}\left(r,\varphi\right)\sigma_{12} \qquad (3.195)$$
$$\gamma_{S,r\varphi,EVZ} = J_{S,31,EVZ}\left(r,\varphi\right)\sigma_{11} + J_{S,32,EVZ}\left(r,\varphi\right)\sigma_{22} + J_{S,33,EVZ}\left(r,\varphi\right)\sigma_{12}$$

Inclusion: Kern:

$$\varepsilon_{K,rr,EVZ} = J_{K,11,EVZ}\left(r,\varphi\right)\sigma_{11} + J_{K,12,EVZ}\left(r,\varphi\right)\sigma_{22} + J_{K,13,EVZ}\left(r,\varphi\right)\sigma_{12}$$
$$\varepsilon_{K,\varphi\varphi,EVZ} = J_{K,21,EVZ}\left(r,\varphi\right)\sigma_{11} + J_{K,22,EVZ}\left(r,\varphi\right)\sigma_{22} + J_{K,23,EVZ}\left(r,\varphi\right)\sigma_{12} \qquad (3.196)$$
$$\gamma_{K,r\varphi,EVZ} = J_{K,31,EVZ}\left(r,\varphi\right)\sigma_{11} + J_{K,32,EVZ}\left(r,\varphi\right)\sigma_{22} + J_{K,33,EVZ}\left(r,\varphi\right)\sigma_{12}$$

3.3.4.2 Displacements

The displacements are determined using the kinematic relationship and integration. The superposition is to be used to each of the strain terms to be integrated.

3.3.4.2 Verschiebungen

Die Verschiebungen werden mithilfe der kinematischen Beziehung und Integration ermittelt, wobei die Superposition auf die einzelnen zu integrierenden Verzerrungsterme angewandt wird.

Radial displacement:

Disc:

Radialverschiebung:

Scheibe:

$$\begin{aligned} u_{S,EVZ} &= \int \varepsilon_{S,rr,EVZ}\,\mathrm{d}r \\ &= \int \left[J_{S,11,EVZ}\left(r,\varphi\right)\sigma_{11} + J_{S,12,EVZ}\left(r,\varphi\right)\sigma_{22} + J_{S,13,EVZ}\left(r,\varphi\right)\sigma_{12} \right]\mathrm{d}r \\ &= \int \varepsilon_{S,r,11,EVZ}\left(r,\varphi\right)\mathrm{d}r + \int \varepsilon_{S,r,12,EVZ}\left(r,\varphi\right)\mathrm{d}r + \int \varepsilon_{S,r,13,EVZ}\left(r,\varphi\right)\mathrm{d}r \\ &= u_{S,11,EVZ} + u_{S,12,EVZ} + u_{S,13,EVZ} \end{aligned} \qquad (3.197)$$

The individual radial displacements $\left(u_{S,11,EVZ}, u_{S,12,EVZ}, u_{S,13,EVZ}\right)$ correspond to the known formulations

Die einzelnen Radialverschiebungen $\left(u_{S,11,EVZ}, u_{S,12,EVZ}, u_{S,13,EVZ}\right)$ entsprechen den bekannten Formulierungen

$$u_{S,11,EVZ} = u_{S,EVZ,N_1}$$
$$u_{S,12,EVZ} = u_{S,EVZ,N_2} \tag{3.198}$$
$$u_{S,13,EVZ} = u_{S,EVZ,N_{12}}$$

and can be found in equations(3.83), (3.102) and (3.160).

und können den Gleichungen (3.83), (3.102) und (3.160) entnommen werden.

Inclusion:

Kern:

$$
\begin{aligned}
u_{K,EVZ} &= \int \varepsilon_{K,rr,EVZ} \mathrm{d}r \\
&= \int \left[J_{K,11,EVZ}(r,\varphi)\sigma_{11} + J_{K,12,EVZ}(r,\varphi)\sigma_{22} + J_{K,13,EVZ}(r,\varphi)\sigma_{12} \right] \mathrm{d}r \\
&= \int \varepsilon_{K,r,11,EVZ}(r,\varphi)\mathrm{d}r + \int \varepsilon_{K,r,12,EVZ}(r,\varphi)\mathrm{d}r + \int \varepsilon_{K,r,13,EVZ}(r,\varphi)\mathrm{d}r \\
&= u_{K,11,EVZ} + u_{K,12,EVZ} + u_{K,13,EVZ}
\end{aligned}
\tag{3.199}
$$

The individual radial displacements $\left(u_{K,11,EVZ}, u_{K,12,EVZ}, u_{K,13,EVZ}\right)$ can be taken also from the already known equations (3.85), (3.104) and (3.162):

Die einzelnen Radialverschiebungen $\left(u_{K,11,EVZ}, u_{K,12,EVZ}, u_{K,13,EVZ}\right)$ können ebenfalls den bereits bekannten Gleichungen (3.85), (3.104) und (3.162) entnommen werden:

$$u_{K,11,EVZ} = u_{K,EVZ,N_1}$$
$$u_{K,12,EVZ} = u_{K,EVZ,N_2} \tag{3.200}$$
$$u_{K,13,EVZ} = u_{K,EVZ,N_{12}}$$

Tangential displacement:

Tangentialverschiebung:

The individual tangential displacements are also already known and can be taken alternatively from the given references.

Die einzelnen Tangentialverschiebungen sind ebenso bereits bekannt und können alternativ den angegebenen Verweisen entnommen werden.

Disc:

Scheibe:

$$
\begin{aligned}
v_{S,EVZ} &= \int \left(r\varepsilon_{S,\varphi\varphi,EVZ} - u_{S,EVZ} \right) \mathrm{d}\varphi \\
&= \int \left[r \left[J_{S,21,EVZ}(r,\varphi)\sigma_{11} + J_{S,22,EVZ}(r,\varphi)\sigma_{22} + J_{S,23,EVZ}(r,\varphi)\sigma_{12} \right] - \left(u_{S,11,EVZ} + u_{S,12,EVZ} + u_{S,13,EVZ} \right) \right] \mathrm{d}\varphi \\
&= \int r\varepsilon_{S,\varphi,EVZ,N_1} - u_{S,EVZ,N_1} + r\varepsilon_{S,\varphi,EVZ,N_2} - u_{S,EVZ,N_2} + r\varepsilon_{S,\varphi,EVZ,N_{12}} - u_{S,EVZ,N_{12}} \mathrm{d}\varphi \\
&= \int r\varepsilon_{S,\varphi,EVZ,N_1} - u_{S,EVZ,N_1} \mathrm{d}\varphi + \int r\varepsilon_{S,\varphi,EVZ,N_2} - u_{S,EVZ,N_2} \mathrm{d}\varphi + \int r\varepsilon_{S,\varphi,EVZ,N_{12}} - u_{S,EVZ,N_{12}} \mathrm{d}\varphi \\
&= v_{S,EVZ,N_1} + v_{S,EVZ,N_2} + v_{S,EVZ,N_{12}}
\end{aligned}
$$

$$\tag{3.201}$$

For the alternative comparison see equations (3.84), (3.103) and (3.161).

Für den alternativen Vergleich siehe Gleichungen (3.84), (3.103) und (3.161)

Inclusion: **Kern:**

$$
\begin{aligned}
v_{K,EVZ} &= \int \left(r\varepsilon_{K,\varphi\varphi,EVZ} - u_{K,EVZ} \right) \mathrm{d}\varphi \\
&= \int r \left[J_{K,21,EVZ}(r,\varphi)\sigma_{11} + J_{K,22,EVZ}(r,\varphi)\sigma_{22} + J_{K,23,EVZ}(r,\varphi)\sigma_{12} \right] - \left(u_{K,EVZ,11} + u_{K,EVZ,12} + u_{K,EVZ,13} \right) \mathrm{d}\varphi \\
&= \int r\varepsilon_{K,\varphi,EVZ,N_1} - u_{K,EVZ,N_1} + r\varepsilon_{K,\varphi,EVZ,N_2} - u_{K,EVZ,N_2} + r\varepsilon_{K,\varphi,EVZ,N_{12}} - u_{K,EVZ,N_{12}} \mathrm{d}\varphi \\
&= \int r\varepsilon_{K,\varphi,EVZ,N_1} - u_{K,EVZ,N_1} \mathrm{d}\varphi + \int r\varepsilon_{K,\varphi,EVZ,N_2} - u_{K,EVZ,N_2} \mathrm{d}\varphi + \int r\varepsilon_{K,\varphi,EVZ,N_{12}} - u_{K,EVZ,N_{12}} \mathrm{d}\varphi \\
&= v_{K,EVZ,N_1} + v_{K,EVZ,N_2} + v_{K,EVZ,N_{12}}
\end{aligned}
$$

$$(3.202)$$

For the alternative comparison see equations (3.86), (3.105) and (3.163).

Für den alternativen Vergleich siehe Gleichungen (3.86), (3.105) und (3.163)

3.3.4.3 Stresses

The formulation of the stresses can be made with the known strains from the equations (3.195), (3.196) and by means of the constitutive relationship:

3.3.4.3 Spannungen

Die Formulierung der Spannungen kann mit den bekannten Dehnungen aus den Gleichungen (3.195), (3.196) und mithilfe der konstitutiven Beziehung erfolgen:

Disc: **Scheibe:**

$$
\sigma_{S,r,EVZ} = \frac{E_1}{1-\mu_1^2}\left(\varepsilon_{S,r,EVZ} + \mu_1\varepsilon_{S,\varphi,EVZ} \right)
$$

$$
\sigma_{S,\varphi,EVZ} = \frac{E_1}{1-\mu_1^2}\left(\mu_1\varepsilon_{S,r,EVZ} + \varepsilon_{S,\varphi,EVZ} \right) \tag{3.203}
$$

$$
\tau_{S,r\varphi,EVZ} = G_1\gamma_{S,r\varphi,EVZ}
$$

Inclusion: **Kern:**

$$
\sigma_{K,r,EVZ} = \frac{E_2}{1-\mu_2^2}\left(\varepsilon_{K,r,EVZ} + \mu_2\varepsilon_{K,\varphi,EVZ} \right)
$$

$$
\sigma_{K,\varphi,EVZ} = \frac{E_2}{1-\mu_2^2}\left(\mu_2\varepsilon_{K,r,EVZ} + \varepsilon_{K,\varphi,EVZ} \right) \tag{3.204}
$$

$$
\tau_{K,r\varphi,EVZ} = G_2\gamma_{K,r\varphi,EVZ}
$$

Alternative

As a control for the stresses, they can be determined by inverting the compliance matrix. This involves a considerable amount of computation, the scope and presentation of which should not be the content of this work. Therefore, the derivation is described below in limited form by means of the matrix calculation. With the inverse of the compliance matrix

Alternative

Als Kontrolle für die Spannungen kann diese durch die Invertierung der Nachgiebigkeitsmatrix ermittelt werden. Damit ist ein beträchtlicher Rechenaufwand verbunden, dessen Umfang und Darstellung nicht Inhalt dieser Arbeit sein soll. Daher wird die Herleitung nachfolgend in, auf die Matrizenrechnung, eingeschränkter Form beschrieben. Die Steifigkeitsmatrix ist die Inverse der Nachgiebigkeitsmatrix

$$\mathbf{J} = \begin{bmatrix} J_{11,EVZ}\left(r,\varphi\right) & J_{12,EVZ}\left(r,\varphi\right) & J_{13,EVZ}\left(r,\varphi\right) \\ J_{21,EVZ}\left(r,\varphi\right) & J_{22,EVZ}\left(r,\varphi\right) & J_{23,EVZ}\left(r,\varphi\right) \\ J_{31,EVZ}\left(r,\varphi\right) & J_{32,EVZ}\left(r,\varphi\right) & J_{33,EVZ}\left(r,\varphi\right) \end{bmatrix}$$

and by means of the determinant   und wird mit der Determinante

$$\det\left(\mathbf{J}\right) = J_{11,EVZ}\left(J_{22,EVZ}J_{33,EVZ} - J_{23,EVZ}J_{32,EVZ}\right)$$
$$-J_{12,EVZ}\left(J_{21,EVZ}J_{33,EVZ} - J_{23,EVZ}J_{31,EVZ}\right)$$
$$+J_{13,EVZ}\left(J_{21,EVZ}J_{32,EVZ} - J_{22,EVZ}J_{31,EVZ}\right)$$

and the adjuncts   und der Adjunkten

$$\mathrm{adj}(\mathbf{J}) = \begin{bmatrix} \left(J_{22,EVZ}J_{33,EVZ} - J_{23,EVZ}J_{32,EVZ}\right) & \left(-J_{12,EVZ}J_{33,EVZ} + J_{32,EVZ}J_{13,EVZ}\right) & \left(J_{12,EVZ}J_{23,EVZ} - J_{22,EVZ}J_{13,EVZ}\right) \\ \left(-J_{21,EVZ}J_{33,EVZ} + J_{31,EVZ}J_{23,EVZ}\right) & \left(J_{11,EVZ}J_{33,EVZ} - J_{31,EVZ}J_{13,EVZ}\right) & \left(-J_{11,EVZ}J_{23,EVZ} + J_{21,EVZ}J_{13,EVZ}\right) \\ \left(J_{21,EVZ}J_{32,EVZ} - J_{31,EVZ}J_{22,EVZ}\right) & \left(-J_{11,EVZ}J_{32,EVZ} + J_{31,EVZ}J_{12,EVZ}\right) & \left(J_{11,EVZ}J_{22,EVZ} - J_{21,EVZ}J_{12,EVZ}\right) \end{bmatrix}$$

the stiffness matrix is calculated: berechnet:

$$\mathbf{K} = \mathbf{J}^{-1} = \frac{1}{\det\left(\mathbf{J}\right)}\mathrm{adj}\left(\mathbf{J}\right) \tag{3.205}$$

Disc: **Scheibe:**

$$\mathbf{K}_S\left(r,\varphi\right)=\left[\mathbf{J}_S\left(r,\varphi\right)\right]^{-1}=\frac{1}{\det\left[\mathbf{J}_S\left(r,\varphi\right)\right]}\,\mathrm{adj}\left[\mathbf{J}_S\left(r,\varphi\right)\right] \qquad (3.206)$$

For the element $K_{S,11,EVZ}$ of the stiff-ness matrix of the disc thus follows:

Für das Element $K_{S,11,EVZ}$ der Steifig-keitsmatrix der Scheibe ergibt sich somit:

$$K_{S,11,EVZ}\left(r,\varphi\right)=\frac{\left(J_{S,22,EVZ}J_{S,33,EVZ}-J_{S,23,EVZ}J_{S,32,EVZ}\right)}{\left[\begin{array}{l}J_{S,11,EVZ}\left(J_{S,22,EVZ}J_{S,33,EVZ}-J_{S,23,EVZ}J_{S,32,EVZ}\right)\\-J_{S,12,EVZ}\left(J_{S,21,EVZ}J_{S,33,EVZ}-J_{S,23,EVZ}J_{S,31,EVZ}\right)\\+J_{S,13,EVZ}\left(J_{S,21,EVZ}J_{S,32,EVZ}-J_{S,22,EVZ}J_{S,31,EVZ}\right)\end{array}\right]}$$

Inclusion: **Kern:**

$$\mathbf{K}_K\left(r,\varphi\right)=\left[\mathbf{J}_K\left(r,\varphi\right)\right]^{-1}=\frac{1}{\det\left[\mathbf{J}_K\left(r,\varphi\right)\right]}\,\mathrm{adj}\left[\mathbf{J}_K\left(r,\varphi\right)\right] \qquad (3.207)$$

For the EVZ, all mechanical quantities for the infinite disc with an elastic inclusion for arbitrarily combined loads (tension, pressure and shear) are given in analytical form.

Für den EVZ sind nun alle mechanischen Größen für die unendliche Scheibe mit elastischem Kern für beliebig kombinierte Belastungen (Zug, Druck und Schub) in analytischer Form gegeben.

3.4 Substitution at the constitutive level

3.4 Substitution auf konstitutiver Ebene

According to [13], the following material parameter relationship between the plane-strain-state (EVZ) and plane-stress-state (ESZ) can be found at the constitutive level.

Nach [13] lässt sich auf konstitutiver Ebene folgende Materialparameterbeziehung zwischen dem EVZ und ESZ finden.

$$\mu_{EVZ} = \frac{\mu_{ESZ}}{1 - \mu_{ESZ}}$$

$$E_{EVZ} = \frac{E_{ESZ}}{1 - \mu_{ESZ}^2}$$

$$(3.208)$$

In this way, the solution for the plane-strain-state (EVZ) can be formulated by substituting the material parameter relationship (3.208) for an existing solution in the plane-stress-state (ESZ). In addition to the shortened formulation of the solution for the plane-strain-state (EVZ) from the already existing solution for the plane-stress-state (ESZ), the substitution at the constitutive level now also permits the verification of the two independently derived solutions of the plane-stress-state (ESZ) and plane-strain-state (EVZ). This verification is shown here for the structure „disc with circular inclusion" in Fig. 3-36. The example for the verification of the disc with core is shown by the following geometrical and mechanical parameters:

Damit lässt sich bei vorhandener Lösung im ESZ durch Substitution der Materialparameterbeziehung (3.208) die Lösung für den EVZ formulieren. Die Substitution auf konstitutiver E-bene erlaubt nun neben der verkürzten Formulierung der Lösung für den EVZ aus der bereits vorhanden Lösung für den ESZ auch die Verifizierung der beiden von einander unabhängig herge-leiteten Lösungen des ESZ und EVZ. Diese Verifizierung wird an dieser Stelle für die Struktur "Scheibe mit Kern" durch Fig. 3-36 nachgewiesen. Das Beispiel für den Nachweis bei der Scheibe mit Kern wird mit den folgenden geo-metrischen und mechanischen Parame-tern geführt:

Radius of inclusion: $r = 10$ mm

Elastic modulus of disc:

$E_1 = 3200$ N/mm²

Poisson ratio of disc: $\mu_1 = 0{,}3$

Elastic modulus of inclusion:

$E_2 = 6400$ N/mm²

Poisson ratio of inclusion: $\mu_2 = 0{,}2$

Stress load: $\sigma_2\left(x,\pm l/2\right) = 2{,}5$ N/mm²

Kernradius: $r = 10$ mm

Elastizitätsmodul-Scheibe:

$E_1 = 3200$ N/mm²

Querkontraktionszahl-Scheibe:

$\mu_1 = 0{,}3$

Elastizitätsmodul-Kern:

$E_2 = 6400$ N/mm²

Querkontraktionszahl-Kern:

$\mu_2 = 0{,}2$

Spannungslast:

$\sigma_2\left(x,\pm l/2\right) = 2{,}5$ N/mm²

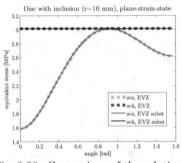 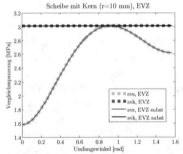

Fig. 3-36: Comparison of the solutions for the plane-strain-state (EVZ) according to derivation and substitution. Vergleich der Lösungen für den EVZ zufolge Herleitung und Substitution.

It clearly shows the identical result for the derived solution and the solution generated by substitution in plane-strain-state (EVZ). By means of this substitution, it is of course also possible to formulate the solutions for the structures "disc with ring" and "disc with inclusion and ring" derived in Volume 1 [67]. In this publication, however, these structures are not considered in detail and thus not further discussed.

Es zeigt sich deutlich das identische Ergebnis für die hergeleitete Lösung und die durch Substitution erzeugte Lösung im EVZ. Mit Hilfe dieser Substitution lassen sich natürlich auch die im 1. Band [67] hergeleiteten Lösungen für die Strukturen "Scheibe mit Ring" und "Scheibe mit Ring und Kern" formulieren. In dieser Arbeit werden diese Strukturen jedoch nicht genauer betrachtet und somit nicht weiter behandelt.

4 Validation of the analytical solution

The validation of the analytical solutions is based on numerical results according to the FEM. For this an example of the plane strain state (EVZ) for the disc with circular inclusion is treated.

The validation of the analytical solution for the **infinite disc** is done by the numerical solution for the **finite disc**. In spite of these different geometric constraints, very good agreement of the solutions are documented. For given sufficient geometrical dimensions, this proves the applicability of the real analytic solution for the infinite disc to the finite disc. Due to the predefined mechanical state „plane strain state" (EVZ), the example is applicable for infinitely extended structures or in states with impaired transverse strain normal to the plane.

As comparison value between the numerical and the analytical solution the Mises comparison stress is used. The choice of the comparison variable is motivated primarily due to the fact that all stress components are taken into account in the comparison stress.

4 Validierung der analytischen Lösung

Die Validierung der analytischen Lösungen erfolgt anhand von numerischen Ergebnissen zufolge der FEM. Dazu wird ein Beispiel des ebenen Verzerrungszustandes (EVZ) für die Scheibe mit Kern behandelt.

Die Validierung der analytischen Lösung für die **unendliche Scheibe** erfolgt anhand der numerischen Lösung für die **endliche Scheibe**. Trotz dieser unterschiedlichen geometrischen Randbedingungen wird eine sehr gute Übereinstimmung der Lösungen dokumentiert. Damit wird nachgewiesen, dass bei ausreichend geometrischen Abmessungen die Anwendbarkeit der reellen analytischen Lösung für die unendliche Scheibe auf die endliche Scheibe gegeben ist. Aufgrund des vorgegebenen mechanischen Zustandes „Ebener Verzerrungszustand" (EVZ) ist das Beispiel für quasi unendlich ausgedehnte Strukturen oder bei Zuständen mit behinderter Querdehnung normal zur Ebene anwendbar.

Als Vergleichsgröße zwischen der numerischen und der analytischen Lösung wird die Mises-Vergleichsspannung herangezogen. Die Wahl dieser Vergleichsgröße ist in erster Linie darin begründet, dass in der Vergleichsspannung alle Spannungskomponenten eine Berücksichtigung finden.

© Springer Nature Switzerland AG 2021
T. Ranz, *Linear Elasticity of Elastic Circular Inclusions Part 2/*
Lineare Elastizitätstheorie bei kreisrunden elastischen Einschlüssen Teil 2,
SpringerBriefs in Applied Sciences and Technology,
https://doi.org/10.1007/978-3-030-72397-2_4

4.1.1 Disc with circular inclusion

A finite square disc with circular inclusion is considered, which according to the plane-strain-state can not deform normal to the disc plane. The plane-strain-state represents the mechanical behavior of an imaginary cut-out strip of a virtually infinite structure. The finite square disc with circular inclusion is loaded on two opposite edges with a line load, which is acting normal to the outer edge. The geometrical and mechanical parameters of the disc with circular inclusion are:

Thickness: $s = 1$ mm
Length: $l = 400$ mm
Radius of inclusion: $r = 10$ mm
Elastic modulus of disc:
$E_1 = 3200$ N/mm^2
Poisson ratio of disc: $\mu_1 = 0{,}3$

Elastic modulus of inclusion:
$E_2 = 6400$ N/mm^2
Poisson ratio of inclusion: $\mu_2 = 0{,}3$

Edge load: $N_2\left(x, \pm l/2\right) = 5$ N/mm
Stress load: $\sigma_2\left(x, \pm l/2\right) = 5$ N/mm^2

4.1.1 Scheibe mit kreirundem Einschluss (Kern)

Es wird eine endliche quadratische Scheibe mit Kern, welche entsprechend dem EVZ sich nicht normal zur Scheibenebene verformen kann, betrachtet. Der EVZ repräsentiert das mechanische Verhalten eines gedachten herausgeschnittenen Streifens einer quasi unendlich ausgedehnten Struktur. Die endliche quadratische Scheibe mit Kern wird an zwei gegenüberliegenden Rändern mit einer normal auf den Außenrand wirkenden Streckenlast belastet. Die geometrischen und mechanischen Parameter der Scheibe mit Kern sind:

Dicke: $s = 1$ mm
Seitenlänge: $l = 400$ mm
Kernradius: $r = 10$ mm
Elastizitätsmodul-Scheibe:
$E_1 = 3200$ N/mm^2
Querkontraktionszahl-Scheibe:
$\mu_1 = 0{,}3$
Elastizitätsmodul-Kern:
$E_2 = 6400$ N/mm^2
Querkontraktionszahl-Kern:
$\mu_2 = 0{,}3$
Randlast: $N_2\left(x, \pm l/2\right) = 5$ N/mm
Spannungslast: $\sigma_2\left(x, \pm l/2\right) = 5$ N/mm^2

The edge load N_2 can also be expressed as a stress load σ_2 on the outer edge, which acts on the slim surfaces $\left(A_S = l \cdot s\right)$ of the cut strip.

Die Randlast N_2 lässt sich ebenso als Spannungslast σ_2 am Außenrand, welche an den schmalen Seitenflächen $\left(A_S = l \cdot s\right)$ des herausgeschnittenen Streifens wirkt, ausdrücken.

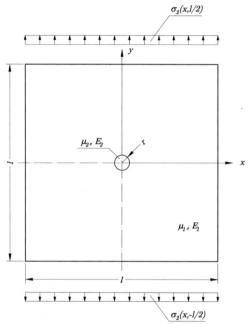

Fig. 4-1: Finite disc (l,l) with inclusion (r) under stress load σ_2 at outer edge. Endliche Scheibe (l,l) mit Kern (r) unter der Spannungslast σ_2 am Außenrand.

For validation, the equivalent stresses along pitch circles with the radii (9.0, 10.0, 11.0, 13.0 and 17.0 mm) in the angular range $\left(0 \leq \varphi \leq \pi/2\right)$ are taken into account. At the diagrams, the numerical solution (FEM) is represented by dashed lines and the real analytic solution by solid lines.

Für die Validierung werden die Vergleichsspannungen entlang von Teilkreisen mit den Radien (9,0; 10,0; 11,0; 13,0 und 17,0 mm) im Winkelbereich $\left(0 \leq \varphi \leq \pi/2\right)$ berücksichtigt. In den Diagrammen wird die numerische Lösung (FEM) durch strichlierte und die reelle analytische Lösung durch ausgezogene Linien dargestellt.

4.1.1.1 FE-Model with shell element

The finite square disc with circular inclusion is mapped in the 1st quadrant by a FE model with symmetric boundary conditions. The FEM mesh of the shell-discretized structure has been refined to convergence.

4.1.1.1 FE-Modell mit Schalenelementen

Die endliche quadratische Scheibe mit Kern wird im 1. Quadranten durch ein FE-Modell mit symmetrischen Randbedingungen abgebildet. Das FEM-Netz der mit Schalenelementen diskretisierten Struktur wurde bis zur Konvergenz verfeinert.

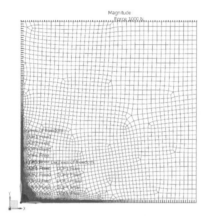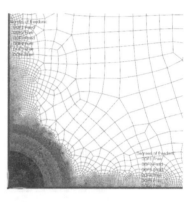

Fig. 4-2: FE-Model with boundary conditions and mesh.
FE-Modell mit Randbedingungen und Netzdiskretisierung.

4.1.1.2 FE-Model – solution

The stresses of the numerical FE solution show in radial direction a continuous and in the tangential direction a discontinuous behavior. This behavior is given due to the change in stiffness between inclusion and disc.

4.1.1.2 FE-Modell – Ergebnis

Die Spannungen der numerischen FE-Lösung zeigen in radialer Richtung ein kontinuierliches und in tangentialer Richtung ein diskontinuierliches Verhalten. Dieses Verhalten ist auf den Steifigkeitssprung zwischen Kern und Scheibe zurückzuführen.

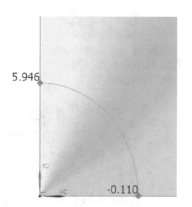
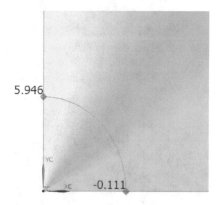

Fig. 4-3: Radial stress (FEM) at inclusion (left) and at disc (right) due to uniaxial loading σ_2. Radialspannung (FEM) im Kern (links) und in der Scheibe (rechts) zufolge einachsialer Belastung σ_2.

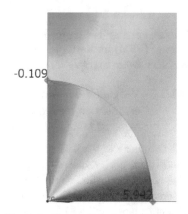

Fig. 4-4: Tangential stress (FEM) at inclusion (left) and at disc (right) due to uniaxial loading σ_2. Tangentialspannung (FEM) im Kern (links) und in der Scheibe (rechts) zufolge einachsialer Belastung σ_2.

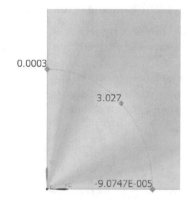

Fig. 4-5: Shear stress (FEM) at inclusion (left) and at disc (right) due to uniaxial loading σ_2. Schubspannung (FEM) im Kern (links) und in der Scheibe (rechts) zufolge einachsialer Belastung σ_2.

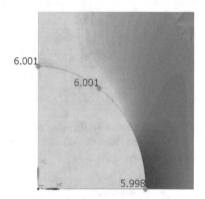
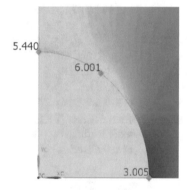

Fig. 4-6: Equivalent stress (FEM) at inclusion (left) and at disc (right) due to uniaxial loading σ_2. Vergleichsspannung (FEM) im Kern (links) und in der Scheibe (rechts) zufolge einachsialer Belastung σ_2.

4.1.1.3 Comparison of numeric- and real-analytical- solution

The graphs of the equivalent stresses (numeric, analytic real) show a high congruence over the entire circumference angle. The equivalent stresses in the inclusion are shown as constant. The stresses in the disc are not constant. The different in stresses between disc and inclusion is clearly described

4.1.1.3 Vergleich der numerischen mit der reellen analytischen Lösung

Die Graphen der Vergleichsspannungen (numerisch, analytisch reell) zeigen eine hohe Übereinstimmung über den gesamten Umfangswinkel. Die Vergleichsspannungen im Kern werden als konstant ausgewiesen und am Berührrand zwischen Scheibe und Kern wird der Spannungssprung eindeutig be-

at the contact edge. With increases of radius, the equivalent stress in the disc approximates the amount of load stress at the outer edge.

schrieben. Mit zunehmendem Radius nähert sich die Vergleichsspannung in der Scheibe betragsmäßig der Spannungslast am Rand.

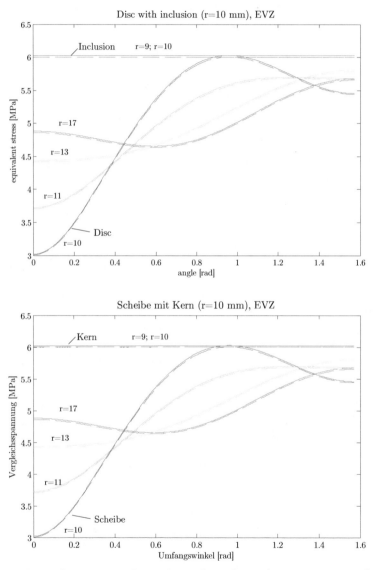

Fig. 4-7: Equivalent stress at disc with circular inclusion (--- numeric, — real)
Vergleichsspannung in der Scheibe mit Kern (--- numerisch, — reell)

5 Summery and outlook

5 Zusammenfassung und Ausblick

This book describes the derivation of the real analytical solutions in plane-strain-state for the „disc with circular inclusion" using the linear elasticity theory in relation to the infinite disc. The solutions take into account the material parameters (elastic modulus and poisson ratio) of the individual structural parts (disc and inclusion) and are formulated for the normal force load and thrust load acting on the infinite edge. They represent a complete and closed solution, which describes the mechanical behaviour separately for each structural part (disc and inclusion).

The solution process consists of the free cutting of the structural parts (disc - inclusion) and the creation of statically indeterminate functions. The displacement compatibility at the cut free contact edge determines the unknown parameters of the indeterminate functions. For this purpose, suitable approaches for the Airy stress functions are to be selected and determined for different indeterminate functions of the substructures. The mechanical quantities (deformations, strains and stresses) of the individual substructures are superposed in accordance with compatibility. This allows the formulating of overall solution for each individual substructure.

Dieses Buch beschreibt die Herleitung der reellen analytischen Lösungen im EVZ für die „Scheibe mit Kern" mithilfe der linearen Elastizitätstheorie in Bezug auf die unendliche Scheibe. Die Lösungen berücksichtigen die Materialparameter (Steifigkeitsmodul und Querkontraktionszahl) der einzelnen Strukturanteile (Scheibe und Kern) und werden für die Normalkraft- und Schubkraftbelastung, welche am unendlichen Rand wirken, formuliert. Sie stellen vollständige und geschlossene Lösungen dar, welche jeden Strukturteil (Scheibe, Kern) separat mechanisch beschreiben.

Der Lösungsprozess besteht aus dem Freischneiden der Strukturteile (Scheibe – Kern) und dem Ansetzen von statisch unbestimmten Funktionen. Mithilfe der Verschiebungskompatibilität an den freigeschnittenen Berührrändern werden die unbekannten Parameter der unbestimmten Funktionen bestimmt. Dazu sind für die unterschiedlichen unbestimmten Funktionen der Teilstrukturen geeignete Ansätze für die Airy'schen Spannungsfunktionen zu wählen und zu bestimmen. Die mechanischen Größen (Verschiebungen, Verzerrungen und Spannungen) der einzelnen Teilstrukturen werden entsprechend der Kompatibilität superponiert und dadurch die Gesamtlösung für jede einzelne Teilstruktur formuliert.

© Springer Nature Switzerland AG 2021
T. Ranz, *Linear Elasticity of Elastic Circular Inclusions Part 2/ Lineare Elastizitätstheorie bei kreisrunden elastischen Einschlüssen Teil 2*, SpringerBriefs in Applied Sciences and Technology, https://doi.org/10.1007/978-3-030-72397-2_5

The solution for the „disc with circular inclusion" requires 3 unknown parameters for the uniaxial load and 2 unknown parameters for the shear load. Due to the high number of terms, they are grouped into constants wherever possible. Nevertheless, partial solution equations are given over several lines result.

The real analytical solutions are compared with the numerical solution (FEM). There is a very high congruence between the real analytic solutions and the numerical solutions, although the latter are solutions of a finite disc. At this point, reference is made to the „de Saint-Vénant's principle", which describes the decrease influence of finite or infinite edge with sufficient distance of the edge to the inclusion at conservative direction of force, see e.g. [9].

After the solution for the „disc with circular inclusion" in plane-strain-state was presented in this book, in a consequent continuation the solution of the „disc with ring and inclusion" for the plane-strain-state is to be formulated in a subsequent work. Based on this, a three-dimensional material model for a composite consisting of a single fiber with a surface layer would be developed.

However, first of all the solution for the „disc with circular inclusion" can

Die Lösung für die „Scheibe mit Kern" benötigt bei der einachsialen Belastung 3 unbekannte Parameter und bei der Schubbelastung 2 unbekannte Parameter. Aufgrund der hohen Anzahl von Termen werden diese nach Möglichkeit zu Konstanten zusammengefasst. Trotzdem ergeben sich teilweise Lösungsgleichungen über mehrere Zeilen.

Die reellen analytischen Lösungen werden mit der numerischen Lösung (FEM) verglichen. Es zeigt sich eine sehr hohe Übereinstimmung zwischen den reellen analytischen Lösungen und den numerischen Lösungen, obwohl letztere Lösungen einer endlichen Scheibe entstammen. An dieser Stelle wird auf das „de Saint-Vénantsche Prinzip" verwiesen, demzufolge und auf unsere Problemstellung bezogen der Einfluss zufolge endlichem oder unendlichem Rand bei genügend großem Abstand des Randes zum Kern und konservativer Kraftrichtung verschwindet, siehe z.B. [9].

Nachdem in diesem Buch die Lösung für die „Scheibe mit Kern" im EVZ vorgestellt wurde, sollte in konsequenter Fortführung auch die Lösung der „Scheibe mit Ring und Kern" für den EVZ in einer nachfolgenden Arbeit mit Hilfe der Substitution auf konstitutiver Ebene formuliert werden.

Darauf aufbauend würde sich ein dreidimensionales Materialmodell für einen Verbundwerkstoff bestehend aus einer einzelnen Faser mit Randschicht entwickeln lassen.

Zunächst jedoch, lässt sich nun mit der Lösung für die „Scheibe mit Kern" ein

now be used to develop a three-dimensional material model for a composite consisting of a single fiber without a surface layer.

A particularly interesting area of application for the solution of the plane-strain-state is the fatigue strength, especially at very high cycle fatigue (VHCF). Due to the now calculable stress peak at elastic inclusions in a base material, these stresses can be approximated even by small amplitude in the collectivization on the very high number of cycles. Their share of the total amount of damage is therefore quantifiable and accessible to the different damage hypotheses.

The solution for the „disc with circular inclusion and ring" for the plane-strain-state is briefly described below in terms of process technology and is analogous to the solution for the plane-stress-state, see [67]. It is a multi-stage solution process. In the first step, the disc with ring is cut free from the inclusion and then the indeterminate functions are to be applied. In the second step, the disc is cut free from the ring under the load of indeterminate functions. For the uniaxial load, a total number of 9 unknown parameters are required, which are determined by the three relationships of the displacement compatibility. Shear loading requires 6 unknown parameters, which are also determined by three relationships of the displacement compatibility.

dreidimensionales Materialmodell für einen Verbundwerkstoff bestehend aus einer einzelnen Faser ohne Randschicht entwickeln.

Ein besonders interessanter Anwendungsbereich für die Lösung des EVZ ist die Ermüdungsfestigkeit, speziell jene bei sehr hohen Schwingspielzahlen (very high cycle fatigue, VHCF). Aufgrund der nun berechenbaren Spannungsspitze bei elastischen Einschlüssen in einem Grundmaterial können diese Beanspruchungen auch von kleiner Amplitude bei der Kollektivierung über die sehr hohe Schwingspielzahl angenähert berücksichtigt werden. Ihr Anteil an der gesamten Schädigungssumme ist somit quantifizierbar und den unterschiedlichen Schädigungshypothesen zugänglich.

Die Lösung für die „Scheibe mit Kern und Ring" für den EVZ wird nachfolgend kurz prozesstechnisch beschrieben und ist analog der Lösung für den ESZ zu führen, siehe [67]. Sie besitzt einen mehrstufigen Lösungsprozess. Im ersten Schritt wird die Scheibe mit Ring vom Kern freigeschnitten und danach sind die unbestimmten Funktionen anzusetzen. Im zweiten Schritt wird unter den Belastungen der unbestimmten Funktionen die Scheibe vom Ring freigeschnitten. Für die einachsiale Belastung sind insgesamt 9 unbekannte Parameter erforderlich, welche durch die dreimalige Formulierung der Verschiebungskompatibilität bestimmt werden. Für die Schubbelastung sind 6 unbekannte Parameter erforderlich, welche ebenso durch drei Verschiebungskompatibilitätsbeziehungen bestimmt werden.

The solutions for the plane-strain-state can be extended to engineering applications, which approximate the structural behavior of components. Such components may be, for example, reinforced concrete parts or tunnel tubes. From a purely mathematical point of view, the derived solutions represent coupled field equations, which would be analogous to other physical problems.

The real analytic solutions can be extended to the time-, temperature-, or humidity-dependent range by substituting the material parameters with functions such as those exemplified in [44], [65] and [66].

The solution process as such can also be applied to approximate solutions of finite discs. Thus, it would be possible to formulate new elements for the FEM with only a few parameters, and it would be possible to calculate structures of composite materials with a significantly lower number of degrees of freedom by means of the FEM.

Die Lösungen für den EVZ lassen sich auf Ingenieuranwendungen, welche das Strukturverhalten von Bauteilen angenähert beschreiben lässt, erweitern. Solche Bauteile können beispielsweise Stahlbetonteile oder Tunnelröhren sein. Rein mathematisch betrachtet stellen die hergeleiteten Lösungen gekoppelte Feldgleichungen dar, welche analog auf andere physikalische Aufgabenstellungen übertragbar wären.

Die reellen analytischen Lösungen lassen sich auf den zeit-, temperatur- oder feuchtigkeitsabhängigen Bereich erweitern, indem die Materialparameter durch Funktionen wie sie beispielhaft in [44], [65] und [66] formuliert sind, substituiert werden.

Der Lösungsprozess als solcher lässt sich auch auf Näherungslösungen von finiten Scheiben anwenden. Damit wäre die Formulierung von neuen Elementen für die FEM mit nur wenigen Parametern möglich und es könnten Strukturen aus Verbundwerkstoffen mit einer wesentlich geringeren Anzahl von Freiheitsgraden mittels der FEM berechnet werden.

6 Bibliography

Literaturverzeichnis

[1]   Argyris, J. H.; Radaj, D.: Parametrische Kerbspannungsuntersuchung am e-
        lastischen Kern. Acta Mechanica Vol. 13, S 303-314, 1972
[2]   Atanackovic, T. M.; Guran, A.: Theory of Elasticity for Scientists and Engi-
        neers. Boston: Birkhäuser 2000
[3]   Bansal, P.; Bansal S. R.: Elastic-Plastic Transition in a Thin-Rotating Disc
        with Inclusion. World Academy of Science, Engineering and Technology 38,
        2008
[4]   Bathe, K. J.: Finite-Elemente-Methoden. Berlin: Springer-Verlag 1986
[5]   Bonnet, M.: Kunststoffe in der Ingenieuranwendung. 1. Auflage. Wiesbaden:
        Vieweg+Teubner 2009
[6]   Castles, R. R.; Mura, T.: The analysis of eigenstrains outside of an ellipsoidal
        inclusion. Journal of Elasticity, Vol. 15, pp 17-34, 1985
[7]   Ernst, G.; Hühne, C.; Rolfes, R.: Micromechanical voxel unit cell for strength
        analysis of fiber reinforced plastics. Conference on Damage in Composite Ma-
        terials, Stuttgart 2006
[8]   Eshelby, J. D.: The determination of the elastic field of an ellipsoidal inclusion, and
        related problems. Proceedings of the Royal Society of London. Series A,
        Mathematical and Physical Sciences, Volume 241, S 376-396, 1957
[9]   Girkmann, K.: Flächentragwerke. 6. Auflage. Wien: Springer-Verlag 1986
[10]   Ghosh, S.; Lee, K.; Raghavan, P.: A multi-level computational model for
        multi-scale damage analysis in composite and porous materials. Int. Journal
        of Solids and Structures, Vol. 38, pp 2335-2385, 2001
[11]   Gladwell, G. M. L.: On Inclusions at a Bi-Material Elastic Interface. Journal
        of Elasticity. Vol. 54, S 27-41, 1999
[12]   Gross, D.; Seelig, T.: Bruchmechanik. 4. Auflage. Berlin: Springer-Verlag
        2007
[13]   Guggenberger, W.: Flächentragwerke, Studienbehelf - Teil 1, Kap. 2 - Schei-
        ben, Institut für Baustatik, Technische Universität Graz, 2019
[14]   Hahn, O.; Kurzok, J. R.; Rohde A.: Untersuchung zur Übertragbarkeit von
        Kennwerten einer punktgeschweißten Einelementprobe auf Mehrelement-
        prüfkörper und Bauteile. Forschungsvereinigung Automobiltechnik e. V. Nr.
        142, 1998
[15]   Hansen, A. C.; Kenik, D. J.; Nelson, E. E.: Multicontinuum Failure Analysis
        of Composites. Presented at ICCM, Edinburgh, UK, 2009
[16]   Hashin, Z.: Analysis of Composite Materials – A Survey. Journal of Applied
        Mechanics, Vol. 50, S 481-505, 1983

© Springer Nature Switzerland AG 2021   95
T. Ranz, *Linear Elasticity of Elastic Circular Inclusions Part 2/*
Lineare Elastizitätstheorie bei kreisrunden elastischen Einschlüssen Teil 2,
SpringerBriefs in Applied Sciences and Technology,
https://doi.org/10.1007/978-3-030-72397-2_6

[17]   Hasselman, D. P. H.; Fulrath, R. M.: Micromechanical Stress Concentrations
        in Two-Phase Brittle-Matrix Ceramic Composites. Journal of The American
        Ceramic Society. Vol. 50, S 399-404, 1967

[18]   Hengst, H.: Beitrag zur Berechnung von Stegblechen mit Sparlöchern. Der
        Stahlbau, Beilage zur Zeitschrift „Die Bautechnik". Jahrgang 15, Heft 17/18,
        S 61-64, 14. August 1942

[19]   Hinton, M. J.; Kaddour, A. S.; Soden, P. D.: Evaluation of failure prediction
        in composite laminates: background to ´part B´of the exercise. Composites
        Science and Technology, Vol. 62, pp 1481-1488, 2002

[20]   Hinton, M. J.; Kaddour, A. S.; Soden, P. D.: Evaluation of failure prediction
        in composite laminates: background to ´part C´of the exercise. Composites
        Science and Technology, Vol. 64, pp 321-327, 2004

[21]   Hinton, M. J.; Kaddour, A. S.: The Second World-Wide Failure Exercise:
        Benchmarking of Failure Criteria under triaxial Stresses for Fiber-Reinforced
        Polymer Composites. 16th International Converence on Composite Materials,
        Kyoto, Japan, 2007

[22]   Hufenbach, W.; Kroll L.: Kerbspannungsanalyse anisotrop faserverstärkter
        Scheiben. Archive of Applied Mechanics. Vol. 62, S 277-290, 1992

[23]   Hütter, A.: Die Spannungsspitzen in gelochten Blechscheiben und Streifen.
        Zeitschrift für angewandte Mathematik und Mechanik. Band 22, S 322-3335,
        1942

[24]   Inglis, C. E.: Stresses in a Plate due to the Presence of Cracks and Sharp
        Corners. Read at the Spring Meetings of the Fifty-fourth Session of the Insti-
        tution of Naval Architects, March 14, 1913

[25]   In-Plane Shear Properties of Unidirectional Fiber/Resin Composite Cylin-
        ders. Military Standard. MIL-STD-375, 1992

[26]   Irwin, G. R.: Onset of Fast Crack Propagation in High Strength Steel. Naval
        Research Laboratory. Washington. D. C., 1956

[27]   Johlitz, M.; Diebels, S.: Effective mechanical behaviour of filled polymers.
        Mechanics of Advanced Materials and Structures, Volume 18, Issue 2, 2011

[28]   Kaczynski A.; Monastyrskyy B.: Thermal stresses in a periodic two-layer
        space with an interface rigid inclusion under uniform heat flow. Acta
        Mechanica. Vol. 203, S 183-195, 2009

[29]   Kaddour, A. et. al: Damage Theories for Fibre-Reinforced Polymer Compos-
        ites: The Third World-Wide Failure Exercise (WWWFE-III). 16th Interna-
        tional Converence on Composite Materials, Kyoto, Japan, 2007

[30]   Kaiser, G.: Die Scheibe mit elliptischem Kern. Archive of Applied Mechanics,
        Vol. 30, S 275-287, 1961

[31]   Kalmykov, Y. B.; Drakin, N. V.; Dubrava, O. L.: Effect of filler size and
        concentration on the physicomechanical properties of a composite polymer
        material. Mech. of Composite Materials, Vol. 25, Issue 2, pp 144-152, 1989

[32]   Kienzler, R.; Schröder, R.: Einführung in die Höhere Festigkeitslehre. Berlin:
        Springer-Verlag 2009

[33] Kim, K.; Sudak, L. J.: Interaction between a radial matrix crack and a three-phase circular inclusion with imperfect interface in plane elasticity. International Journal of Fracture, Volume 131, pp 155-172, 2005

[34] Kirsch, G.: Die Theorie der Elastizität und die Bedürfnisse der Festigkeitslehre. Zeitschrift des Vereines deutscher Ingenieure, 42, pp 797–807, 1898.

[35] Knight; M. G.; et. al.: A study of the interaction between a propagating crack and an uncoated/coated elastic inclusion using the BE technique. Int. Journal of Fracture, Vol. 114, pp 47-61, 2002

[36] Krenkel, W.: Keramische Verbundwerkstoffe. Weinheim: WILEY-VCH Verlag GmbH & Co. KGaA 2003

[37] Kroll, L. et. al.: Analytical Models for Stress and Failure Analysis of Notched Hyprid Composites. 21st International Congress on Theoretical and Applied Mechanics, Warschau, 15.-21. August 2004

[38] Kröner, E.: Die Spannungsfunktionen der dreidimensionalen isotropen Elastizitätstheorie. Zeitschrift für Physik, Bd. 139, S 175-188, 1954

[39] Lauke, B.; Schüller, T.: Calculation of stress concentration caused by a coated particle in polymer matrix to determine adhesion strength at the interface. Composites Science and Technology. Vol. 62, S 1965-1978, 2002

[40] Leon, A.: Über die Spannungsstörungen durch Kerben und Tellen und über die Spannungsverteilung in Verbundkörpern. Österreichische Wochenschrift für den öffentlichen Baudienst - Mitteilung aus dem mechanisch-technischen Laboratorium der k. k. technischen Hochschule. 1. Teil, Heft 43, S 770-776, 2. Teil, Heft 44, S 783-790, 1908

[41] Leon, A.: Zur Theorie der Verbundkörper. Armierter Beton. II Jahrgang, S 343-351, 1909

[42] Lepper M. et. al.: Kerbspannungsanalyse textilbewehrter Holzkonstruktionen mittels analytischer Berechnungsverfahren. Textilbeton – 1. Fachkolloquium der Sonderforschungsbereiche 528 und 532 in Aachen, 2001

[43] Li, Z. R.; Lim, C. W.; He, H. L.: Stress concentration around a nano-scale spherical cavity in elastic media: effect of surface stress. European Journal of Mechanics A/Solids. Vol. 25, S 260-270, 2006

[44] Lion, A.: Thermomechanik von Elastomeren. Habilitation. Berichte des Instituts für Mechanik (Bericht 1/2000). Gesamthochschul-Bibliothek Kassel

[45] Meder, G.: Zur exakten und näherungsweisen Berechnung unidirektionalverstärkter Kunststoffe. Materialwissenschaft und Werkstofftechnik, Vol. 12, S 366-374, 1981

[46] Menges, G.; Gitschner, H.-W.: Einfluß korrosiver Medien auf das Verhalten von glasfaserverstärkten Verbundwerkstoffen. Materials and Corrosion, Vol. 31, S 693-702, 1980

[47] Mishra, D.; et. al: Elliptical Inclusion Problem in antiplane Piezoelectricity: Stress concentrations and energy release rates. International Journal of Fracture, Volume 179, pp 213-220, 2013

[48] Moser, K.: Faser-Kunststoff-Verbund: Entwurfs- und Berechnungsgrundlagen. Düsseldorf: VDI-Verlag 1992

[49]   Muskhelishvili, N. I.: Some basic problems of the mathematical theory of
       elasticity. Reprint of the second English edition. Leyden: Noordhoff Interna-
       tional Publishing 1977

[50]   Mußchelischwili, N. I.: Einige Grundaufgaben zur mathematischen Elastiz-
       itätstheorie. München: Carl Hanser Verlag 1971

[51]   Munz, M.: Evidence for a three-zone interphase with complex elastic-plastic
       behaviour: nanoindentation study of an epoxy/thermoplastic composite.
       Journal of Physics D: Applied Physics, Volume 39, Number 18, 2006

[52]   Müller, W. H.: Mathematical vs. Experimental Stress Analysis of Inhomoge-
       neities in Solids. Journal de Physique IV. Colloque C1, supplement au Jour-
       nal de Physique III, Vol. 6. 1996

[53]   Neuber, H.: Kerbspannungslehre. 3. Auflage. Berlin: Springer-Verlag 1985

[54]   Pak, Y. E.; Mishra, D.; Yoo, S. H.: Closed-form solution fo a coated circular
       inclusion under uniaxial tension. Acta Mechanica 223, 937-951, 2012

[55]   Papanicolaou, G. C.; Theocaris, P. S.; Spathis, G. D.: Adhesion efficiency be-
       tween phases in fibre-reinforced polymers by means of the concept of bound-
       ary interphase. Colloid and Polymer Science, Volume 258, Issue 11, pp 1231-
       1237, 1980

[56]   Paris, P. C.; Gomez, M. P.; Anderson, W. P.: A Rational Analytic Theory of Fa-
       tigue. The Trend In Engineering, Vol. 13, S 9-14, 1961

[57]   Pineda, E. J.: Progressive damage and failure modeling in notched laminated fiber
       reinforced composites. Int. Journal of Fracture, Vol.158, Issue 2, pp 125-143, 2009

[58]   Pöschl, T. Über eine partikuläre Lösung des biharmonsichen Problems für
       den Außenraum der Ellipse. Mathematische Zeitschrift, Volume 11, Issue 1-2,
       pp 89-96, 1921

[59]   Pöschl, T.: Die Verwendung von Spannungsfunktionen beim statischen Scha-
       lenproblem. Zeitschrift für technische Physik. Nr. 8, S 216-222, 1921

[60]   Puck, A.: Festigkeitsanalyse von Faser-Matrix-Laminaten - Modelle für die
       Praxis. München: Hanser 1996

[61]   Radaj, D.: Fatigue Assessment of Spot Welds by Approximated Local Stress
       Parameters. International Journal of Fracture. Vol. 102, S L3-L8, 2000

[62]   Radaj, D.; Kandel, W.: Kerbspannungen am Last tragenden elastischen
       Kern. Forschung im Ingenieurwesen. Vol: 39, S 1-12, 1973

[63]   Rahman, M.: A Rigid Elliptical Disc-Inclusion, in an Elastic Solid, Subjected
       to a Polynomial Normal Shift. Journal of Elasticity. Vol. 66, S 207-235, 2002

[64]   Rahman, M.: The Stress-Intensity Factor for a Rigid Circular Disc-Inclusion,
       under Axial Translation, Embedded into an Elastic Cylinder. International
       Journal of Fracture, Vol. 110, S 23-27, 2001

[65]   Ranz, T.: Elementare Materialmodelle der Linearen Viskoelastizität im Zeit-
       bereich. Beiträge zur Materialtheorie (Heft 5/07). München: Institut für Me-
       chanik 2007. ISSN 1862-5703

[66]   Ranz, T.: Viskoelastische Materialmodell für Holz. Experimente, Model-
       lierung und Simulation. VDI-Verlag 2009

[67] Ranz, T.: Lineare Elastizitätstheorie bei kreisrunden elastischen Einschlüssen. Shaker-Verlag 2014

[68] Rapp, H.: Inhomogene Balken unter Termperatur-, Zug- und reiner Beigebeanspruchung. TU-München, Dissertation 1988

[69] Richter, M.; Lepenies, I.; Zastrau, B. W.: Modellierung des Materialverhaltens von Textilbeton mittels repräsentativer Volumenelemente. Aktuelle Beiträge aus Baustatik und Comp. Mechanics, Universität der Bundeswehr, Selbstverlag, 2003

[70] Rudolph, J.; Weiß, E.; Forster, M.: Modeling of welded joints for design against fatigue. Engineering with Computers. Vol. 19, S 142-151, 2003

[71] Sadd, M. H.: Elasticity, Theory, Applications, and Numerics. Oxford: Elsevier 2009

[72] Sadowsky, M. A.; Sternberg, E.: Stress Concentration Around a Triaxial Ellipsoidal Cavity. Journal of Applied Mechanics. S 149-157, JUNE 1949

[73] Sanadi, A. R.; Subramanian, R. V.; Manoranja, V. S.: The Interphasial Regions in Interlayer Fiber Composites. Polymer Composites, Volume 12, Issue 6, pages 377–383, 1991

[74] Schaefer, H.: Die Spannungsfunktion des dreidimensionalen Kontinuums und des elastischen Körpers. ZAMM, Bd. 33, S 356-362, 1953

[75] Schwabe, F.: Einspieluntersuchungen von Verbundwerkstoffen mit periodischer Mikrostruktur. RWTH-Aachen, Dissertation 2000

[76] Schürmann, H.: Konstruieren mit Faser-Kunststoff-Verbunden. Berlin: Springer-Verlag 2005

[77] Selvadurai, A. P. S.: An inclusion at a bi-material elastic interface. Journal of Engineering Mathematics, Vol. 37, S 155-170, 2000

[78] Shen, L.; Li, J.: Homogenization of a fibre/sphere with an inhomogeneous interphase for the effective elastic moduli of composites. Proc. R. Soc. A, vol. 461, no. 2057, pp 1475-1504, 2005

[79] Soden, P. D.; Kaddour, A. S.; Hinton, M. J.: Recommendations for designers and researchers resulting from the world-wide failure exercise. Composites Science and Technology, Vol. 64, pp 589-604, 2004

[80] Strubecker, K.: Airysche Spannungsfunktionen und isotrope Differentialgeometrie. Mathematische Zeitschrift. Vol. 78, S 189-198, 1962

[81] Szabó, I.: Höhere Technische Mechanik. Korrigierter Nachdruck der 5. Auflage. Berlin: Springer-Verlag 1977

[82] Timpe, A.: Die Airysche Funktion für den Ellipsenring. Mathematische Zeitschrift, Vol. 17, Nr. 1, S 189-205, 1923

[83] VDI 2014 Blatt 3: Entwicklung von Bauteilen aus Faser-Kunststoff-Verbund. Berechnungen. Berlin: Beuth Verlag GmbH 2006

[84] Weber, C.: Halbebene mit Kreisbogenkerbe. Zeitschrift für angewandte Mathematik und Mechanik. Band 21, S 230-232, 1941

[85] Weber, C.: Spannungsverteilung in Blechen mit mehreren kreisrunden Löchern. Zeitschrift für angewandte Mathematik und Mechanik. Band 2, S 267-273, 1922

[86]   Weber, C.: Zur Spannungserhöhung bei gelochten gezogenen Streifen. Zeitschrift für angewandte Mathematik und Mechanik. Band 21, S 252, 1941

[87]   Wolf, K.: Beiträge zur ebenen Elastizitätstheorie. Zeitschrift für technische Physik. Nr. 8, S 209-216, 1921

[88]   Zhang S.: On Debonding of Thin Films. International Journal of Fracture. Vol. 119, S L9-L14, 2003

[89]   Zhang S.: On the Interface Crack Between Two Elastic Layers Under General Edge Loads. International Journal of Fracture. Vol. 112, S L27-L32, 2001

[90]   Zhang S.: Stress Intensity Factors for Spot Welds Joining Sheets of Unequal Thickness. International Journal of Fracture. Vol. 122, S L119-L124, 2003

[91]   Zhang, S.: Fracture mechanics solutions to spot welds. International Journal of Fracture. Vol. 112, S 247-274, 2001

[92]   Zhang, S.: Stress intensities derived from stresses around a spot weld. International Journal of Fracture. Vol. 99, S 239-257, 1999

[93]   Zienkiewicz, O. C.: Methode der finiten Elemente. 2. Auflage. München: Carl Hanser Verlag, 1984

Printed in the United States
by Baker & Taylor Publisher Services